T0093755

Scientific Journeys

"What a joy to accompany Fred Dylla on his *Scientific Journeys*! This engaging and insightful collection of essays, drawn from a lifetime in science, takes you through a wide-ranging landscape of research, with fascinating excursions into history, politics, art, and personalities along the way."

—Colin Norman, *Former News Editor, Science magazine*

"A delightful collection of short essays; each has a surprise, something memorable about science, or people, or his personal trajectory, or strategy, or the startling importance of science in the world. So many thrills in a little book!"

—John Mather, *Nobel Prize (physics), 2006*

"*Scientific Journeys* is a gem. These delightful essays illuminate the fascinating lives of a diverse group of scientists and engineers and the unpredictable, rewarding paths of discovery they followed to create our modern world. Start your journey with Hildegard von Bingen, a nun who left a scientific and musical legacy more than 900 years ago, and end with Fred Dylla's optimistic epilogue that science will provide a cure for the COVID-19 pandemic and bring the world into a better place."

—Madeleine Jacobs, *Former Editor in Chief, Chemical & Engineering News and Executive Director and CEO, American Chemical Society*

"This is an interesting, thoughtful, and well-written tour of the world of a physicist, from his education at MIT, mentored by some of the greatest physicists of the day, to his eventual position as CEO of the American Institute of Physics."

—Mark Cardillo, *Executive Director of the Camille and Henry Dreyfus Foundation*

"Fred Dylla delightfully shares the arc of his career from building his own laser as a boy to leading the American Institute of Physics in this collection of engaging essays. I enjoyed reading them all and learned many things along the way."

—Kevin Marvel, *Executive Director of the American Astronomical Society*

"With an insider's perspective, *Scientific Journeys* ably shows how science and engineering innovations emerge from a complex network of policy deliberations, inspired ideas, dedicated work, adroit management of research, and timely sharing of results."

—Phillip Schewe, *Author of The Grid, a history of how society uses electricity, and Maverick Genius, a biography of the scientist and writer Freeman Dyson*

"We are fortunate that Fred Dylla has devoted the effort to share some of the interesting experiences and perspectives from his distinguished career in physics. They are well written, insightful, and even entertaining, especially for those of us who have also chosen to be physicists."

—Bob Doering, *Director of Exploratory Research, Technology and Manufacturing Group, Texas Instruments*

"Fred's extraordinary career and range as an essayist are nicely summed up in Chap. 26 by his observation: '…physicists like to say they are interested in everything … I was not immune to this.' While greatly enjoying each essay, I was reminded how, as a conference organizer, I used to put Fred at the end to ensure that no delegates drifted away early! Sadly, no such inspirational book existed when I started out, but I am confident that *Scientific Journeys* will do much to motivate the next generation. I shall certainly be buying copies for young friends and grandchildren."

—Robert Campbell, *Former President of Blackwell Publishing*

H. Frederick Dylla

Scientific Journeys

A Physicist Explores the Culture, History and Personalities of Science

With a Foreword by Rush D. Holt

 Springer

H. Frederick Dylla
Lewes, DE, USA

Foreword by
Rush D. Holt
Washington, DC, USA

ISBN 978-3-030-55799-7   ISBN 978-3-030-55800-0   (eBook)
https://doi.org/10.1007/978-3-030-55800-0

© Springer Nature Switzerland AG 2020
This work is subject to copyright. All rights are reserved by the Publisher, whether the whole or part of the material is concerned, specifically the rights of translation, reprinting, reuse of illustrations, recitation, broadcasting, reproduction on microfilms or in any other physical way, and transmission or information storage and retrieval, electronic adaptation, computer software, or by similar or dissimilar methodology now known or hereafter developed.
The use of general descriptive names, registered names, trademarks, service marks, etc. in this publication does not imply, even in the absence of a specific statement, that such names are exempt from the relevant protective laws and regulations and therefore free for general use.
The publisher, the authors and the editors are safe to assume that the advice and information in this book are believed to be true and accurate at the date of publication. Neither the publisher nor the authors or the editors give a warranty, expressed or implied, with respect to the material contained herein or for any errors or omissions that may have been made. The publisher remains neutral with regard to jurisdictional claims in published maps and institutional affiliations.

Frontispiece: Propagation of the guide laser at the Gemini North Observatory, Hilo, HI. *Credit* Gemini Observatory/AURA

Artwork: The images on the title pages introducing each part of this book are white-line woodcuts by the author

This Springer imprint is published by the registered company Springer Nature Switzerland AG
The registered company address is: Gewerbestrasse 11, 6330 Cham, Switzerland

To Professors John G. King and Rainer Weiss for their inspiring teaching and
dedication to scientific research
and to Linda for unfailing encouragement

Foreword

In this volume, we have the wide-ranging thoughts and observations of Fred Dylla, an accomplished physicist with an engineer's fascination for gadgets, a historian's long perspective, an artist's aesthetic eye, and a teacher's passion for sharing ideas. Throughout his varied career as a scientist, builder, administrator, publisher, and craftsman, his curiosity has been his foremost characteristic and his ability to see the connection between apparently disparate things is his greatest skill. Occasionally throughout his career, and especially in recent years when he headed the large scholarly consortium and scientific publisher, the American Institute of Physics, he has recorded his thoughts in remarkably varied essays on personalities, patents, policies, publishing, and artistic productions. Here, he examines the roots and growth of innovation in examples from Bell Laboratories, Edison Electric Light Company, and cubist painter Georges Braque. He considers the essential place of publishing in science, the epochal intellectual technique for learning how the world works. He shows the human enrichment and practical benefits that are derived from wise investments in scientific research, as well as the waste resulting from a failure to embrace appropriate technologies.

As the reader will see in these essays, Fred Dylla's all-consuming drive to extend his understanding of the world began young. When he was only a freshman in high school, scientists from an industrial physics laboratory, recognizing his technical precocity, gave him a large, expensive ruby crystal so that he could build a laser, with which he began biology experiments. He went on to the Massachusetts Institute of Technology (MIT) where, in his undergraduate and doctoral work, he performed experiments in acoustics, low-temperature physics, biophysics, and surface physics. His mentor while studying all of these fields was John G. King, one of MIT's most outstanding and highly acclaimed teachers of physics. Along the way, he worked with Amar Bose, who built a gigantic acoustics corporation, and Rainer Weiss, who received the 2017 Nobel Prize for detecting cosmic gravitational waves. Later, as the head of research groups at national laboratories he worked with gases heated to millions of degrees and with high-energy electron beams driven by super-cold cavity accelerators. As an officer of scientific societies and as Chief Executive Officer of the American Institute of Physics (AIP), he advanced science and science policy. In my own career as a physics researcher, teacher, policy maker, and elected official, I have interacted with Fred frequently over the years, and I admire his scientific skill, his personal consensus-building in research, business, and policy, and his passion to communicate and to educate.

I first met Fred Dylla when we both worked at Princeton University's Princeton Plasma Physics Laboratory in the late 1980s and early 1990s. PPPL, as it was known, had grown over the previous few decades from a small group of scientists who wanted to apply the new knowledge about the behavior of hot gases and magnetic fields in stars to the clever idea of making fusion energy on earth. The clever idea proposed to use containers with no walls that is made only of magnetic fields and filled with plasma gas at temperatures of millions of degrees, to fuse hydrogen, deuterium, and tritium and release energy in a process that would be economically and environmentally attractive. By the time we were there, PPPL was a major national laboratory of the U.S. Department of Energy. Both Fred and I, like the thousand or so other scientists, engineers, and technicians who worked at PPPL, were attracted by these large and beautiful research projects, the enormous sophistication of the fusion challenge, and the momentous societal benefit that could be foreseen. Fred worked on techniques to heat the hydrogen gases and to measure their behavior at high temperatures. Along the way, he arranged for national industrial partners to adopt plasma technologies that had been developed at PPPL. This program of technology transfer continues today. Fred also helped develop safety procedures at PPPL to assure the Department of Energy, the university, and the community of the safe handling of tritium gas. He and his advisory committee showed that significant releases of tritium would be very unlikely, and that the risks from tritium, although it is radioactive, would be very small even compared to existing radioactivity in our natural and human-made environment. One amusing and effective moment came when Fred pointed out to inquiring non-scientists that tritium had already been used for years in public places, in sealed tubes in luminous exit signs like the ones in the room where they were sitting.

Most noteworthy for me was Fred's work on science education. For some years before I arrived as assistant director of PPPL, Fred had helped to organize the laboratories' scientists to invite interesting researchers from every kind of science to speak at a weekly *Science on Saturday* program. Intended to acquaint high school students with actual scientific research, the program attracted parents and grandparents, as well. When Fred left the laboratory several years later, I assumed the oversight of that program, which has continued ever since. Some parents have continued to come regularly, long after their children have graduated for school, and now some of the previous high school students are bringing their own children to *Science on Saturday*.

That community program for science education illustrated Fred's continual interest in what science and society owe each other. There is a tendency for scientifically trained and non-scientific communities to separate from each other intellectually, even to think of each other as ignorant or illiterate. The British chemist, novelist, and government official, C. P. Snow, stimulated an international discussion sixty years ago when he wrote *The Two Cultures and the Scientific Revolution* [1], saying that science and the humanities seem to employ separate languages and cultural views. Snow's analysis is oversimplified. There are many fault lines of comprehension and appreciation running through and between the scientific and non-scientific cultures. However, the mutual incomprehension becomes much more than a cultural curiosity when it affects public support for the science the public needs for wise policy decisions, and when scientists fail to show the public the importance of evidence-based thinking in the day-to-day fulfillment of a citizen's democratic responsibility.

Fred left PPPL to help build the Jefferson Lab in Newport News, Virginia, and set up to study nuclear physics with high-energy electron beams, and medical and other applications of high-power lasers. Through it all, he continued to build industrial partnerships and to engage in public education. He built local and international collaborations using the laboratories' technologies.

He participated in education activities for students at all levels, and he spoke widely with policy makers and the public about the need to support scientific research. Many citizens may question the relevance of research such as conducted at an electron beam facility. Fred convincingly showed that, in addition to practical spin-offs, such as medical treatments (from which Fred himself has benefited), such work contributes to the growing web of understanding about our world that is part of civilization's centuries-long progress toward a better, more equitable, and more sustainable quality of life. As I moved into public life as a member of the US Congress, elected by the people of central New Jersey, I occasionally saw Fred and appreciated his views on the state of science and on successful techniques for the education of both scientists and the public.

One important policy matter that Fred and I have worked on together, when I was in Congress and later when I headed the American Association for the Advancement of Science, is scientific publishing. As the former head of the major publishing organization, AIP, Fred understands that good communication of scientific findings is essential for science to thrive and to benefit society—indeed for science to work at all. He also understands that good scientific communication is not frictionless and cost-free, even in the modern digital age when so much information seems free and instantaneous. If science is to generate knowledge that is more and more reliable, researchers must make their findings openly available so that other researchers can critique, build on, refine, or refute those findings. However, well-edited, well-reviewed journals are necessary to minimize inaccurate and accidentally or deliberately misleading reports. In recent years, some people say that making scientific articles free to all readers would facilitate the scientific process and would be fairer to taxpayers and other funders of research. Some governments are proposing to require that research with government funding be published free to all readers, a departure from the principal model for scientific journals. Mandating publishing at zero price does not magically produce publishing with zero cost. Someone must pay in one way or another for the maintenance and enforcement of accuracy and quality. Mandated models of scientific communication, if constructed without a good understanding of how scientific communication can be supported, could have an adverse effect on the progress of science.

In 2009, Representative Bart Gordon, Chair of the U.S. House Committee on Science and Technology, asked Fred to convene a Scholarly Publishing Roundtable of publishers, librarians, and university research administrators to consider the issue of public access to government-funded research. The recommendations of Fred's group became the basis for legislation enacted to facilitate accessible publishing of federally funded research. Fred then worked to form a nonprofit corporation supported by the publishing community to implement the legislation now known as CHORUS.

In this collection of essays, Fred Dylla touches on these areas where our careers have intersected and many more. He has been to many places and has gotten to know some of the world's scientific greats, both through personal contact as a student and as an international scientific leader, and through archival sojourns and museum study. The reader will find much food for thought.

Washington, DC, USA

October 2019

Rush D. Holt

Executive Director Emeritus, American

Association for the Advancement of Science

Washington, DC, USA

Reference

1. C.P. Snow, *The Two Cultures and the Scientific Revolution*, Cambridge University Press, Cambridge, U.K. (1959)

Introduction

I consider myself very fortunate to have been born shortly after the cessation of World War II and to have grown up in 1950's America. The nation and world were changed considerably by that terrible cataclysm that cost tens of millions of lives and left huge swaths of Europe, Russia, and the Far East in ruins. The American continent escaped the devastation of the battlefields, but the gargantuan war effort had transformed American industry and American social life in just four years. Millions of returning American soldiers took advantage of the offer from the US Congress to attend college and vocational schools with the tuition cost paid by the government. This legislation, the so-called GI Bill, led to the most educated generation in the nation's history with more than 2 million obtaining their college degrees within 10 years of the war's end. Over its first decade, the legislation cost the nation more than 4 billion dollars, but the estimated return on this investment over the next generation was a 100-fold [1]. I was one of the so-called baby boomer generations that benefited from the postwar infusion of an educated workforce and expanding economy that appeared unchallenged and unlimited at the time. The largess in the American continent was generously shared for rebuilding the devastated economies of the European continent and Japan—which led to both Germany and Japan becoming major industrial powers of the world within three decades after the war.

The warring nations' investment in science and technology during the war had obvious consequences. It is often said that the allied effort to develop and deploy radar won the war and the Manhattan Project's successful development of the first atomic weapons ended the war, with Japan's surrender in the Pacific theater. The German efforts to develop rocket and jet-powered weapons at Peenemünde led to the successful development of missiles that guarded the stalemate peace of the Cold War, and to the race for the first moon landing in 1969.

The small committee of very talented scientist–statesmen that steered the research effort in the USA during and immediately after the war (led by MIT's Vannevar Bush and Harvard's James Conant) laid the groundwork for considerable and sustained funding of basic science by the US government after the war starting with the founding of the Office of Naval Research in 1946 and the National Science Foundation in 1950. These agencies became models for funding scientific research in the USA (and other countries) and have since seen near-continuous growth year after year, despite fluctuations due to downturns in the national economy and political support that varied with how well incoming political factions understood the long-term value of government investment in science.

I grew up in 1950's America with the second national awakening of the value of science. After the Soviets launched the world's first artificial satellite in 1957, there was considerable angst about the nation being dangerously behind in the "space race." The angst was converted into the passage of the National Defense Education Act in 1958. This enabled significant new investments, not only in space technology but in nearly all forms of scientific research and scientific and engineering education. I became what was known as a "Sputnik Kid," convinced that I was going to be scientist and with never a thought about another career path—despite being eight years old at the time of Sputnik's launch. I was joined by many other Sputnik Kids in my generation. We built and launched small rockets, took apart and re-assembled radio receivers and transmitters, and collected all kinds of rocks and minerals—hoping for some spectacular crystals or perhaps some radioactive ores that would phosphoresce on a bookshelf at night. I devoured any sort of science magazines I could get my hands on, such as *Popular Science* and *Scientific American*. (Millennial readers will find this latter form of information transfer archaic given the advent of the Web and Google.)

Despite this limited exposure to the scientific literature, I knew I was going to be a scientist. I did not know what kind of scientist but did not particularly care. All of that changed with the invention of the laser in 1960. I read and re-read an article in *Popular Science* called "*The Incredible Ruby Ray*." From the time I finished the article, I was convinced of two things: I needed to build my own laser, and I was going to be a physicist. Both came true.

This book is a collection of essays that were written after I had spent four decades in various capacities as a working physicist. My trajectory of training to be a physicist and then working as a physicist was never easy, but it was always satisfying. I shared my career with hundreds of unforgettable colleagues and occasional less collegial beings who I felt were impediments to my view of a sensible path forward. I learned from both types of encounters.

The first phase of my science education occurred in a middle school and high school in New Jersey where I was fortunate to be exposed to teachers that let me plow ahead with self-studies on almost scientific subject that interested me. One of the essays in this book (Chap. 16) provides a window for that period in my life and illustrates the value of generous mentors for nourishing enquiring minds. I filled my toolbox for doing science with the good fortune of spending eight years at MIT, gaining three diplomas in physics in three separate subjects. I was fortunate to have a thesis advisor for that entire period who felt strongly that scientists should change their field of study every five years so that they do not get into a scientific rut. You will have an opportunity to read about this remarkable mentor in the essay written about Prof. John G. King (Chap. 17).

I added something to my college education that few scientists indulge but should, given the importance of good communications and analytic skills to any vocation. During my undergraduate years, I included literature as my second area of study. For the rest of my career, this investment paid off immensely. At the time I just needed a night off from physics or math problem sets, but more importantly, the plays and novels I read taught me to understand that science and engineering need the humanities for these endeavors to be most useful for our survival and sustenance.

After obtaining my Ph.D. in biophysics, I spent the next 33 years in two U.S. National Laboratories, first as a plasma physicist and machine builder at Princeton University's Princeton Plasma Physics Laboratory and then as an accelerator physicist and another machine builder at Jefferson Lab, in Newport News, VA. You will find several essays in this book where I have woven life lessons from these laboratories into the narrative.

In 2007, I embarked on my last formal career stop. From 2007 to 2015, I was Executive Director of the American Institute of Physics. AIP is a federation of ten physics-based societies serving more than 125,000 scientists with scientific publications, and related education, government, and public outreach services. For me, it was the job of a lifetime. It gave me an astounding opportunity to view science worldwide and meet and interact with thousands of scientists. I had left the laboratory and direct employment as a scientist, but I found the trade-off worthwhile.

Most of the essays in this book were written while I was at the helm of AIP. On the day I started working there, I launched a weekly essay for AIP members and staff to note important developments that could affect the progress of science or commemorate an historical or cultural event that would cross-connect with science and enrich both endeavors. I have grouped these essays in five parts. The opening Part I entitled *"Signposts"* is vignettes for a number of personalities or events in science that affected my world view of science and culture. In Part II, *"Mentors and Milestones,"* I highlight the science and engaging personalities that intersected my career from building lasers, to magnetic fusion research, to particle accelerators, to materials science.

Part III, *"Science Policy Matters,"* describes my forays in explaining the importance of science to a whole range of audiences: from my children and fellow students, to the general public, to informed decision makers, and to legislators who handle the purse strings for scientific research. Part IV, *"Communicating Science,"* explores a particular form of science communication to which I devoted much time in my professional life—the scientific journal. Such journals are needed for scientists to present their work to their peers. Journal publication is a complicated and expensive business that is undergoing significant stress due to the burgeoning volume of content and the need for rapid incorporation of new technology to deliver the content. I describe pragmatic solutions for stress relievers that I helped put in place to keep the scientific journal sustainable, while other forms of initially print communications (i.e., newspapers and news periodicals) are largely ghosts of their original incarnations.

In Part V *"Art and Science,"* I hope you find pure fun. I have always enjoyed the interplay between art and science. My two daughters (Kim and Sarah) are accomplished artists who would have made fine scientists, but they thought their father was a little too crazy and instead sought out (successful) careers in the arts. They taught me to appreciate the arts, they are wonderful critics of my attempts to generate art, and they continue to show me the value of integrating the appreciation of art and science. The collection of essays in this last part illustrates these points.

Please dear reader, do not feel that you have to read this collection in any order. Pick and choose essays that may entice you at the moment. For those interested, I have offered footnotes and references for further reading on many of the included topics.

During my entire career, I was also fortunate to be exposed to many talented scientists and humanists who enriched my life. I mention and thank just a few in the *Acknowledgements* section at the end of the book.

Author with an inspiring bronze for all scientists at the Einstein Museum in Bern, SW *Credit* Courtesy of Linda B. Dylla

2020   H. Frederick Dylla

Reference

1. Glenn C. Altschuler and Stuart M. Blumin, *The GI Bill: A new deal for veterans*, Oxford University Press, Oxford, UK, 2009.

Contents

Part IV   Communicating Science

Part V   Art and Science

About the Author

H. Frederick Dylla is Executive Director Emeritus of the American Institute of Physics. He has spent 50 years as an enquiring physicist helping to design and build scientific facilities for astronomy, fusion energy, particle and nuclear physics, and medical and materials research. Along the way, he enjoyed sharing his love of science and the arts with students, colleagues, family, and friends.

He lives in Lewes, Delaware, with his wife Linda.

Part I
Signposts

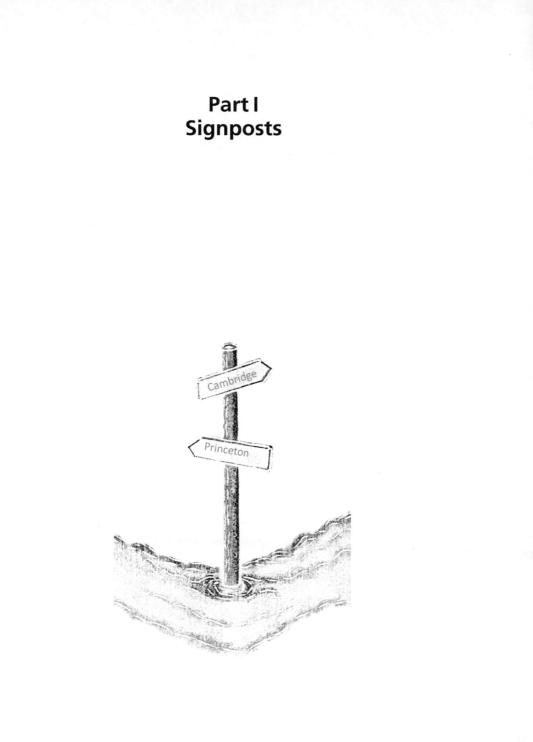

1

Literature and Legacy Flow Along the Rhine

The month of October is known to everyone in the publishing business as the time to head to Germany for the annual Frankfurt Book Fair. This event is believed to be one of the oldest trade fairs in the world, having started more than 500 years ago, when every book was handmade. The German Rhineland is indeed an appropriate location for celebrating the written word.

Situated about an hour's drive from Frankfurt, nestling along the west bank of the Rhine River, is the small village of Bingen. With part of my scientific career dedicated to scientific journal publishing, I often attended the Frankfurt Book Fair. To recover from this heady atmosphere of more than 2500 publishers and 2.5 million books, I frequently headed to this town for some needed rest and refreshment. Today, Bingen is famous for its fine Riesling wines, but its true allure lies in certain remarkable historical events that transpired on its soil from the Roman era onward; the region was witness to many defining developments of European history.

More than 900 years ago, Bingen was also the home of an influential woman and a bright signpost of scholarship, "Hildegard von Bingen." Hildegard began her journey to fame at the age eight when she was sent to a convent. By the time she died in 1179, at the age of 81, she had established three convents on her own, written three major theological studies and one treatise on the natural sciences, invented an alphabet, established a new form of music with over 70 original compositions, and ensured her legacy by making sure archival versions of her life's work were produced and distributed from the Mediterranean to the British Isles.

© Springer Nature Switzerland AG 2020
H. Frederick. Dylla, *Scientific Journeys*,
https://doi.org/10.1007/978-3-030-55800-0_1

Fig. 1.1 Sculpture of Hildegard von Bingen at the Eibingen Abbey Church of St. Hildegard in Rüdesheim am Rhein, Germany by Karlhienz Oswald. *Credit* Photograph by Gerda Arendt, CC-BY-SA 3.0

It is humbling to contemplate what an exceptional individual Hildegard was. Remarkably, most of her work has been preserved; the convents that she founded were not so fortunate. In Middle Europe, where the strategic Rhine Valley was a scene of near constant conflict, only traces of the original majestic structures remain on the Rhine hills. But her works, which she first transcribed onto wax tablets, were then further transcribed by the copy machines of the day—monks with quill pens scribing on parchment—and those copies have remained intact for nearly 1000 years. After the invention of the Gutenberg press, of course, copies of her work proliferated. The Bingen town museum has an archival 1533 edition of Hildegard's work *Physica*, [1] containing her studies on plants, medicine, minerals, and metals. Visitors can stroll through a nearby garden displaying the plants she cultivated for their medicinal qualities.

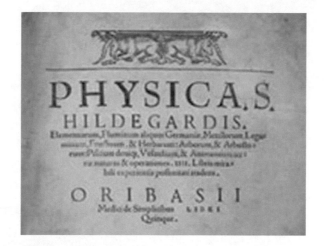

Fig. 1.2 First print of *Physica* 1533. *Credit* Museum of the River

What Hildegard discovered and taught about medicinal botany no doubt gave comfort to the many who visited her convent. The contemporary visitor will often be treated to perhaps her most enjoyable bequests to humankind—her music. During one Bingen visit, I heard two superbly simple, breathtaking performances of Hildegard's music performed by small women's choirs, whose voices were amplified by the perfect acoustics of a stone chapel called the Rochuskapelle.

Of particular interest to the scientific community—the International Astronomical Union has named one of the thousands of minor planets that orbit the Sun after Saint Hildegard of Bingen: number "898 Hildegard."

Reference

1. Hildegard von Bingen, *Physica*, 1533; http://www.landderhildegard.de/her-life/healer-with-natural-remedies/

2

Invention and Discovery: Fleming and Edison

When it comes to scientific progress, many factors come into play: necessity, opportunity, resources, timing, and even luck. Yet perhaps most vital to scientific advancement are the drive, ingenuity, and influence of brilliant individuals. Learning about an invention's history or a discovery gives insights into its genesis. The learning experience is enlivened by the inventors' and innovators' personalities.

Fig. 2.1 John Ambrose Fleming. *Credit* Elliot and Fry, *Page's Engineering Weekly*, London, 1906

© Springer Nature Switzerland AG 2020
H. Frederick. Dylla, *Scientific Journeys*,
https://doi.org/10.1007/978-3-030-55800-0_2

While conducting research for an article I was writing about the birth of electronics [1], I became intrigued by John Ambrose Fleming, a remarkable inventor about whom I knew little despite the impact of his invention. Fleming invented the vacuum diode in 1904. This was the first efficient detector of radio waves and thus launched the first half-century of electronics built around the radio industry.

Fleming began a long career as the first professor of electrical engineering at University College London, but he also worked for the Edison Electric Light Company and the Marconi Company. Because Fleming was comfortable working both in the scientific/academic environment as a researcher and professor and in the technological/commercial realm as an engineer and business consultant, he became a pioneer of the kind of cross-sector collaboration that is so often fruitful.

Fig. 2.2 Thomas Alva Edison, 1922. *Credit* Library of Congress

Fleming's invention of the vacuum diode was based on the work of the world's most prolific inventor, Thomas Edison, who in 1882 introduced a wire into one of his newly invented incandescent lamps. Edison noticed that a current was transmitted to this extra wire from the lamp's filament. Just three years after he had demonstrated the first practical incandescent lamp,

Edison designed, financed, installed, and began operating a fully functional electrical power system in lower Manhattan.

In this short period, he figured out how to manufacture light bulbs in quantity, design and build steam power generators, and lay more than 14 miles of electrical cable to distribute power to customers. Edison built efficient electric motors and a whole array of ancillary components that contributed to the power system's functionality—switches, lamp holders, fuses, power meters, and so forth.

In addition, Edison financed the whole project himself, successfully marketing the new energy source against the established energy providers (gas companies) and fought a corrupt city hall (Tammany Hall). The entrenched interests would not have bet on Edison's original venture of the incandescent lamp, but less than 20 years later, more than 2000 operating power stations

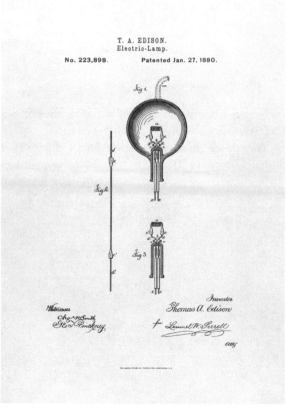

Fig. 2.3 Edison's U.S. Patent on his incandescent lamp. *Credit* Smithsonian Institution, Washington, DC

had been built in the US, and General Electric (the company that Edison founded), Westinghouse, and Western Electric were major manufacturers of electrical products. Harold Evans's book *They Made America: From the Steam Engine to the Search Engine* [2] celebrates the inventor and takes particular notice of Edison's talents.

The science frontiers and engineering challenges of today's highly interconnected world normally require talented teams of people working collectively. In this post-industrial age, any significant impact by a lone inventor or a small entrepreneurial team may seem unlikely. However, in my lifetime I have witnessed James Watson and Francis Crick's discovery of the structure of DNA unleash the new field of molecular biology, Norman Borlaug's development of wheat strains help feed millions across the globe, and Bill Gates' software system and resulting mega-company, Microsoft, make the personal computer a household commodity. Members of the scientific community and their professional societies need to celebrate the impact of the dedicated entrepreneur and help nourish the environment that breeds such entrepreneurs.

References

1. H.F. Dylla and Steven T. Corneliussen, *Journal of Vacuum Science and Technology* **A23**, 1244, (2005); https://doi.org/10.1116/1.1881652
2. Harold Evans, *They Made America: From the Steam Engine to the Search Engine* (Little, Brown and Company, 2004)

3

Rutherford's Nuclear World

History provides teaching moments for many fields of study. For the sciences, I have found that exploring how a major discovery played out, with its usual fits and starts, is an enjoyable and satisfying learning experience. Eureka moments are hard won. Subsequently, written textbook entries for a discovery often lay out the minimal logical path from stated problem to found solution. Such methodical reconstructions, however, strip away the real-life pain and reward and make the path less interesting to me.

I had the opportunity at an annual meeting of physics teachers [1] in 2011 to put one of the most important discoveries in physics in its historical context. The year 2011 was the centenary of Ernest Rutherford's discovery of the atom's nucleus [2], a structure much smaller than the atom itself (by a factor of more than 10,000), but nevertheless containing nearly all the atomic mass. Even to a bright elementary student, none of this is surprising today. But in the early 1900s, the discovery was profound. Very little was known about the atom except for its approximate size and that it contained some negatively charged particles we now call electrons.

Rutherford's discovery of the nucleus was conducted over a five-year period that began in 1908 when his students Hans Geiger and Ernest Marsden started measuring the trajectories of alpha particles emanating from a natural radioactive source directed onto a thin foil of gold. As the researchers expected, most of the particles traversed the foil with only a small deflection from their forward path. To the experimenters' and Rutherford's surprise, however, a few were deflected through large angles. After puzzling for about a year over how this could happen, Rutherford suggested that a massive, charged body lay deep within the atom. His analysis of the experiment has

© Springer Nature Switzerland AG 2020
H. Frederick. Dylla, *Scientific Journeys*,
https://doi.org/10.1007/978-3-030-55800-0_3

held up to this day, though he and his students were careful to do many follow-up experiments before they declared success.

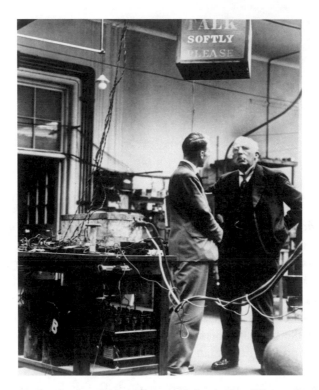

Fig. 3.1 Ernest Rutherford and his colleague John A. Ratcliff standing near a sensitive instrument that picked up audible signals from a charged particle detector. Hence the sign above, "Talk Softly Please". *Credit* Photograph by C.E. Wynn-Williams, courtesy of AIP Emilio Segrè Visual Archives

Rutherford and his students are credited with two milestone developments that the international physics community memorialized in 2011: the discovery of the atomic nucleus and the invention of the particle scattering experiment. Particle scattering remains, a century later, the primary means of investigating the subatomic world. Rutherford's work launched the fields of nuclear physics and later particle physics, which have given us our basic understanding of the Universe's structure from subatomic to cosmic dimensions.

Having spent a not insignificant fraction of my career as a physicist designing and building particle accelerators, there is one particular talk given by Rutherford that really intrigued me. In 1927, just as he came to the end of his position as president of Britain's Royal Society, Rutherford turned his attention to the need for, and benefit to science of a much more efficient way of accelerating particles to high energy than had been possible over the first quarter of the twentieth century. He hypothesized that a configuration that could accelerate charged particles to over a million volts of kinetic energy in the space of a soapbox would have huge ramifications for our understanding of the makeup of matter. Rutherford's "Million Volts in a Soapbox" speech launched a century-long quest to develop accelerators with higher and higher energies. Rutherford's call for a device where physicists could turn up the energy with one knob and the quantity of accelerated particles with another knob would give science the flexible tool needed to fully investigate the subatomic world. What was not known at the time was that this same tool would also end up being a window into the physics of the Universe.

Rutherford's investigations in the first decade of the 1900s that led to his seminal discovery of the nucleus in 1911 used natural sources of alpha and beta rays to provide energetic probing particles. These sources had obvious limitations, in particular, fixed and relatively low energies, in addition to being weak sources of incident particles. His team moved to using cosmic rays, which routinely shower the Earth from extraterrestrial sources such as the Sun and cataclysmic events in the Universe. The highest energy particles ever detected have come from cosmic ray measurements, but the intensities are too weak for most investigations.

At the time of Rutherford's call for a "million volts in a soapbox," the first inventions of particle accelerators were underway. In 1924, the first device to use radiofrequency fields in an evacuated tube to accelerate charged particles was invented by Ising [3]. The year 1932 saw a remarkable series of advances with Rutherford's protégé at Cambridge's Cavendish Lab, where Cockcroft and Walton accelerated protons through over half a million volts [4], and at Berkeley, where E.O. Lawrence demonstrated his first cyclotron, breaking the million volt barrier with a device that could be held in your hand [5]. In that same year, Cockcroft and Walton's machine discovered the other particle hiding in the nucleus, the neutron, harder to spot because it was uncharged [6], and the electron's dopelgänger, the positron, with its positive charge, was also discovered [7]. From that momentous year through to the end of the twentieth century, a family tree of accelerators were designed, built, and operated as one of the most important tools of physics [8]. Each generation of a particular accelerator type would remain on the cutting edge of utility for about 20 years, before becoming obsolete when a new technology came along to enable an order of magnitude jump in energy or intensity.

This evolution continued until the completion and operation of the Large Hadron Collider (LHC) at CERN [9], with its design energy of 8 trillion volts, as the crowning achievement to date of Rutherford's 1927 request. These devices have long since outgrown Rutherford's soapbox. The CERN LHC is contained in a circular tunnel of circumference 27 km that passes under the Jura mountains, through parts of Switzerland and France. The LHC was designed by its creators with one outstanding experiment in mind—to find the so called "Higgs" particle. This particle's existence had been predicted by a group of theoretical physicists, including Peter Higgs from the University of Edinburgh in the mid 1960s, as the mediator of fields within the nucleus accounting for the mass of the proton and neutron—and hence the mass of all matter. Two large teams of physicists using detectors designed and constructed over a decade announced the discovery of the Higgs particle in 2012. That discovery was the subject of the 2013 Nobel Prize in Physics. This award continued a tradition that saw one-third of all Nobel Prizes in Physics over the period of 1939–2013 being dependent on or influenced by measurements using particle accelerators. (See Chap. 23 for more on the LHC.)

New physics discoveries are the obvious end goals of international megaprojects in science such as the LHC. What is often overlooked are the significant complementary benefits stemming from these collaborations. An analysis by the U.S. Department of Energy in 2010 [10] showed that over 30,000 particle accelerators were then operating worldwide. Most of these machines are in use for processing and inspecting materials or for medical diagnosis and therapy. When the LHC was in its design stage in the early 1990s, the team's information scientists, led by Tim Berners-Lee, introduced the world to a new communication tool for the burgeoning internet—the World Wide Web. This tool was designed to ship terabytes of data among the world-wide collaboration of over 10,000 scientists who use CERN's data. From that original application, the Web has become the world's most important form of daily communication and commerce—a very nice pay-off for an investment in the quest for a quirky sub-atomic particle.

Accelerator physicists are still at work designing and bench testing a variety of schemes to move beyond the energies achieved with the LHC. Achieving substantial jumps in energies with extrapolations of currently envisioned hardware is impractical. However, at least one more step with conventional accelerators may be built by international consortia such as CERN, and that could reach approximately ten times the energy of the LHC. The challenge to scientists and engineers is to cap the size and cost to be less than ten times the size and cost of the LHC. Potential scientific solutions that may disrupt the straightforward evolution of existing accelerator technologies are the use of powerful lasers that have recently demonstrated particle accelerations through over a billion volts in a space not much bigger than a soapbox [11]. As we approach the centennial of Rutherford's 1927 challenge, we have moved his goal a thousand-fold to stay on the frontiers of physics.

References

1. A presentation I gave at the Winter 2011 meeting of the American Association of Physics Teachers, Jacksonville, Fl, January 8–12, 2011, entitled: *"Ernest Rutherford and the accelerator: a million volts in a soapbox"* can be found at: https://www.aapt.org/Conferences/wm2011/upload/Rutherford-and-the-Accelerator_final-PDF.pdf
2. The American Institute of Physics produced an informative website on Rutherford *"Rutherford's Nuclear World"* on the occasion of the centenary of his famous 1911 discovery of the nucleus: https://history.aip.org/history/exhibits/rutherford/
3. George Ising, Ark. Mat. Astron. Fys. **18**, 45 (1925)
4. J.D. Cockcroft and E.T.S. Walton, Proc. Roy. Soc. London, Ser **A136**, 619 (1932)
5. E.O. Lawrence and M.S. Livingston, Phys. Rev. **37**, 1707 (1931)
6. James Chadwick, Proc. Roy. Soc. London **A136**, 692 (1932)
7. C.D. Anderson, Phys. Rev. **43**, 491 (1933)
8. The original plot showing the evolution of attained energies by particle accelerators was drawn by E.O. Lawrence's colleague Milton Livingston and published in his book: M.S. Livingston, *High Energy Accelerators,* Interscience, New York (1954); there have many subsequent updates since the original. The following article shows a 21st Century update from G. Krafft at Jefferson Lab (Newport News, VA) in H.F. Dylla, Journal of Vacuum Science & Technology A **21**, S25 (2003): https://doi.org/10.1116/1.1599891
9. The CERN website is a panoply of resources on the physics done at CERN and descriptions of the complex of accelerators including the largest, the Large Hadron Collider: https://home.cern/science/accelerators/large-hadron-collider
10. A useful and not overly technical summary of the applications of particle accelerators to a wide variety of fields including basic and applied research, medicine, environment, energy, and defense is given in this summary report from a workshop sponsored by the U.S. Department of Energy in October, 2009 entitled *Accelerators for America's Future*: https://science.osti.gov/-/media/hep/pdf/accelerator-rd-stewardship/Report.pdf?la=en&hash=4C255E1ED19B65387F6A21910FDD5C16E9FC2A36
11. V. Malka, Physics of Plasmas **19**, 055501 (2012); https://doi.org/10.1063/1.3695389

4

Fueling Science for War and Peace

One man more than any other set the course for American twentieth century science. He was the key scientific advisor during the Roosevelt Administration.

Fig. 4.1 Vannevar Bush, ca. 1940–44. *Credit* U.S. Office for Emergency Management

Vannevar Bush was a Ph.D. MIT engineer who coordinated U.S. scientific research during World War II as director of the Office of Scientific Research and Development. Among his many accomplishments before the war was the invention of one of the first practical analog computers—the differential analyzer—used to solve complex differential equations. Just as the war was

© Springer Nature Switzerland AG 2020
H. Frederick. Dylla, *Scientific Journeys*,
https://doi.org/10.1007/978-3-030-55800-0_4

ending, Bush's experience led him to speculate on the need for what is now called IT-information technology—an enormously better ability to manipulate, share, and capitalize on human knowledge and data. His 1945 article, "*As We May Think*," [1] which appeared in *The Atlantic* magazine, is still available to read. One idea described in this article outlines a device he called a "memex" that could compress, store, and provide access to all forms of one's written communications. His ideas for such a device are often cited as the foundation for the hypertext coding of information. Decades later, like-minded researchers demonstrated Vannevar Bush's prescience by inventing first the Internet and then, at CERN, the World Wide Web. By then we had the benefit of physics-derived technological advances, such as microelectronics, high-density storage, and optoelectronics—all of which were needed to enable our present near-instantaneous access to much of the world's written and visual information.

But Vannevar Bush is remembered even more for how we *go about* achieving scientific and technological advances. At President Franklin D. Roosevelt's request, he produced a widely read and cited government report, "*Science: The Endless Frontier*," [2] which became important for establishing the road map for subsequent development of U.S. scientific enterprise. The report's recommendations also sowed the seed for science investment made by nearly all of the developed nations in the second half of the twentieth century. Bush, in addition to being a remarkable engineer and administrator, was a superb writer. A few excerpts from the report reveal that many of the seemingly new prospects for science in the twenty-first century actually go back two-thirds of a century:

From President Roosevelt's letter requesting the report: "*New frontiers of the mind are before us, and if they are pioneered with the same vision, boldness, and drive with which we have waged this war we can create a fuller and more fruitful employment and a fuller and more fruitful life.*"

From Bush's transmittal letter for the report to the President: "*The pioneer spirit is still vigorous within this nation. Science offers a largely unexplored hinterland for the pioneer who has the tools for his task. The rewards of such exploration both for the Nation and the individual are great. Scientific progress is one essential key to our security as a nation, to our better health, to more jobs, to a higher standard of living, and to our cultural progress.*"

This wartime report had as its context science's huge role in the war effort, but also had as its purpose the establishment of a postwar vision for science in peacetime, especially basic research. It contained an opening section that carried the heading "*Scientific Progress is Essential*." That section began:

Progress in the war against disease depends upon a flow of new scientific knowledge. New products, new industries, and more jobs require continuous additions to knowledge of the laws of nature, and the application of that knowledge to practical purposes. Similarly, our defense against aggression demands new knowledge so that we can develop new and improved weapons. This essential, new knowledge can be obtained only through basic scientific research. Science can be effective in the national welfare only as a member of a team, whether the conditions be peace or war. But without scientific progress no amount of achievement in other directions can insure our health, prosperity, and security as a nation in the modern world.

So how does this science benefit happen? Bush's report recognized the importance of the university in growing the human capital that could sustain the research enterprise:

How do we increase this scientific capital? First, we must have plenty of men and women trained in science, for upon them depends both the creation of new knowledge and its application to practical purposes. Second, we must strengthen the centers of basic research which are principally the colleges, universities, and research institutes. These institutions provide the environment which is most conducive to the creation of new scientific knowledge and least under pressure for immediate, tangible results. With some notable exceptions, most research in industry and Government involves application of existing scientific knowledge to practical problems. It is only the colleges, universities, and a few research institutes that devote most of their research efforts to expanding the frontiers of knowledge.

This science-focused vision of Bush was largely followed by the U.S. and much of the rest of the developed world during the recovery from World War II. Shortly after the publication of Bush's report, the U.S. Office of Naval Research was founded to fund basic and applied research of interest to the naval environment. Five years later in 1950, Bush's vision of a dedicated government agency for broadly funding basic research in many fields came to life with the founding of the U.S. National Science Foundation. The actions that followed Bush's science vision and have spread worldwide since the publication of *Science: The Endless Frontier* remain well-established and proven.

It is fitting to end this brief homage to Vannevar Bush, with the closing words from his 1945 article in *The Atlantic*, which still ring true:

The applications of science have built man a well-supplied house, and are teaching him to live healthily therein. They have enabled him to throw masses of people against one another with cruel weapons. They may yet allow him truly to encompass the great record and to grow in the wisdom of race experience. He may perish in conflict before he learns to wield that record for his true good. Yet, in the application of science to the needs and desires of man, it would seem to be a singularly unfortunate stage at which to terminate the process, or to lose hope as to the outcome.

References

1. Vannevar Bush, "*As we may think*", The Atlantic, July 1945; https://www.theatlantic.com/magazine/archive/1945/07/as-we-may-think/303881/
2. Vannevar Bush, *Science the Endless Frontier*, U.S. Government Report available at: https://www.nsf.gov/about/history/vbush1945.htm

5

Shelter Island's Famous Physicists

History has never been a dull or irrelevant subject for me. Particularly for science, I find that learning about how a scientific concept, discovery, or invention actually transpired is a far more engaging way of seeing how science really works. The script that ends up in science textbooks—hypothesis, analysis, and confirmation—rarely plays out in real life, nor is it as interesting as the real-life version with its pitfalls and wrong turns, and those rare eureka moments of the world's best scientists and engineers.

I find there is often a teaching moment connected to the examination of what happened at a particular time and place. A trip to Long Island, New York, presented such an opportunity. One of the most famous physics conferences occurred on June 2–4, 1947, in a small resort hotel, the Ram's Head Inn on Shelter Island off the far eastern tip of Long Island. The "Shelter Island Conference" has obtained almost iconic status in the history of physics for several reasons: the timing, who attended, and what was accomplished in such a short period of time. By 1947, the United States was just recovering from the end of World War II. The U.S. physics community had been transformed by its essential service to the Allied war effort, and by the infusion of world-renowned physicists who found safe harbor from Nazi persecution in the U.S. The conference was born from the efforts of three scientists: Duncan MacInnes, an electrochemist from the Rockefeller Institute; Frank Jewett, a former head of Bell Laboratories and then president of the National Academy of Science; and Karl Darrow, a Bell Labs physicist who also served as secretary of the American Physical Society. These gentlemen wanted a small conference in a secluded location, because even by the 1940s, typical scientific

© Springer Nature Switzerland AG 2020
H. Frederick. Dylla, *Scientific Journeys*,
https://doi.org/10.1007/978-3-030-55800-0_5

venue meetings were getting too large and this was stifling personal interaction. They wanted to model the conference on the famous Solvay Congresses held in Europe between the World Wars, where the foundations of the "new physics," both quantum mechanics and relativity, were debated.

Fig. 5.1 From the left: Lamb, Pais, Wheeler, Feynman, Feshbach, and Schwinger talking at the Shelter Island Conference, June 1947. *Courtesy* of AIP Emilio Segrè Visual Archives

The timing could not have been better. This turned out to be one of the first physics conferences held after the end of World War II. There was pent-up demand to discuss the subatomic world without the cloak of necessary censorship on sensitive topics related to defense. The chosen conference topic, "The Foundations of Quantum Mechanics," became a bridge in time and space, linking the foundational work in quantum mechanics accomplished in the 1920s and 30s, largely done in Europe, and impending contributions that would stem from the revitalized, multinational U.S. physics community.

Twenty-four American scientists attended the conference, including those with already well-established careers and accomplishments (Robert Oppenheimer, Hans Bethe, and Isidor Rabi), as well as the young prodigies who would go on to transform quantum mechanics (Richard Feynman and Julian Schwinger). The attendees heard about recent discoveries—later recognized by Nobel Prizes—including Willis Lamb's and Robert Retherford's discovery of a slight anomaly in spectral emissions from hydrogen that would become a key driver for the development of the theory since christened "quantum electrodynamics" or "QED." Feynman presented his first ideas on this new theory, now accepted as one of the most precise descriptions of the behavior of electrons and atoms. By 1970, Feynman remarked to one of his biographers, Jagdish Mehra, that Shelter Island was the most important conference he had ever attended. Oppenheimer voiced a similar opinion.

Postwar transportation figured in the success of the conference. As many of the participants headed out to Shelter Island on a bus from Manhattan, it was temporarily detoured to a small Long Island town so that the physicists could share a celebratory meal for their service to the nation during the war. On his train ride home, Hans Bethe worked out some key calculations showing that QED could match the very precise measurements of what is still known as the "Lamb shift."

The Shelter Island Conference [1, 2] had significant influence on both the experimental and theoretical developments of quantum physics. Solvay had moved to Shelter Island. Organizers of modern scientific conferences should note that the entire bill for the conference was $850! Even with factoring in a half century of inflation, its return on investment would be difficult to match.

References

1. Silvan Schweber, "*A Short History of Shelter Island* I." In R. Jackiw, N. Khuri, S. Weinberg, E. Witten. *Shelter Island II: Proceedings of the 1983 Shelter Island Conference on Quantum Field Theory and the Fundamental Problems of Physics.* MIT Press, (1985)
2. American Physical Society, "*The Shelter Island Conference, June 2–4, 1947*", in "*This Month in Physics History*", APS News, June 2000; https://www.aps.org/publications/apsnews/200006/history.cfm

6

Feynman's Scientific Integrity

Scientific integrity has periodically been in the news for the last two decades. The issue resurfaced with vehemence due to the controversy over hacked e-mail messages from groups of climate scientists in 2009 [1]. This incident along with the occasion of similar rare events that are raised to public scrutiny have planted scientific integrity deeply in question for some members of the public.

Arguably the fundamental issue boils down to something described in a comment from a colorful physicist and teacher: the late Richard Feynman. He spoke of "a kind of scientific integrity, a principle of scientific thought that corresponds to a kind of utter honesty—a kind of leaning over backwards. For example, if you're doing an experiment, you should report everything that you think might make it invalid—not only what you think is right about it: other causes that could possibly explain your results; and things you thought of that you've eliminated by some other experiment, and how they worked.... Details that could throw doubt on your interpretation must be given, if you know them.... If you make a theory, for example, and advertise it, or put it out, then you must also put down all the facts that disagree with it, as well as those that agree with it."

This Feynman quotation from his lecture "Cargo Cult Science" [2] reappeared in a 2009 letter to the editor to the *Wall Street Journal* [3] from someone who sees it as a reminder of what climate scientists have forgotten. What I think, however—despite the lapses revealed in the hacked messages—is that it shows what climate scientists have remembered.

© Springer Nature Switzerland AG 2020
H. Frederick. Dylla, *Scientific Journeys*,
https://doi.org/10.1007/978-3-030-55800-0_6

Science works the same in all its fields. If scientific integrity in general weren't assured to a very high degree—in the laboratory, in scientific publications, and in the practical consequences to which science leads—then neither the critics of climate scientists nor the rest of us could be using transistors, lasers, optical fibers, or pharmaceuticals. We couldn't fly with only a one-in-a-million chance of mishap, our bridges and skyscrapers could fall down, and polio would still be feared.

Science is by its very nature an exploratory, trial-and-error venture that is also—sooner or later in every case—a self-correcting exercise. Ukrainian agronomist Trofim Lysenko's [4] failed agricultural theories of the 1930s and 1940s and, more recently, such concepts as polywater, cold fusion, and human clones are examples of scientific pronouncements that were eventually proven wrong or fraudulent by the step-by-step process of examination, review, and repetition.

It's true that scientists are human and that the scientific enterprise can suffer from the frailties of any other human endeavor. I would rather think that scientists are less corrupted than are people in other professions by jealousies, excessive ego, and the desire for fame and fortune. But these human faults do also affect scientists, which mean that science sometimes suffers. Nevertheless, science recovers quickly because of its well-proven correction mechanisms that apply universally across disciplinary, political, and cultural boundaries.

The on-going attack on the integrity of climate science is based on the proposition that this particular field somehow operates with a special, deep disrespect for the skepticism principle that Feynman advocated in the comment quoted above. I state emphatically that no scientific endeavor survives without a healthy respect for skepticism and constant questioning of the evidence. This is how science advances and discoveries are made. And that's exactly why, unfortunately, the dire predictions made by climate scientists are becoming reality.

Fig. 6.1 Richard Feynman teaching at Caltech. *Courtesy* of AIP Emilio Segrè Visual Archives, *Physics Today* Collection and the California Institute of Technology

References

1. For a summary of the controversy and its immediate effects, see David Biello, *"Negating 'Climategate': Copenhagen talks and climate science surviving stolen e-mail controversy"*, *Scientific American*, Feb. 1, 2010; https://www.scientificamerican.com/article/negating-climategate/
2. R.P. Feynman, *Cargo Cult Science*, remarks he made at Caltech's 1974 Commencement Address; http://calteches.library.caltech.edu/51/2/CargoCult.htm
3. Letter to the Editor, *"Feynman on Scientific Integrity"* *The Wall Street Journal*, Dec. 2, 2009; https://www.wsj.com/articles/SB10001424052748703939404574567864090764366

4. An excellent account of Trofim Lysenko's damage to Soviet science in the early 20th century and an unfortunate re-emergence of his praise in 21st Century Russia is given by Sam Kean in "*The Soviet Era's Deadliest Scientist Is Regaining Popularity in Russia*", The Atlantic, December 19, 2017; https://www.theatlantic.com/science/archive/2017/12/trofim-lysenko-soviet-union-russia/548786/

7

Calculating from Memory

The most recent segment of my scientific career has been devoted to the field of science communications—from highly specialized journal articles written for the scientific community, to lay-language communications crafted for the public to highlight scientific developments. On a long flight home after attending a conference on scientific publishing in the U.K., I was struck by the overlap of several related impressions from my trip. My thoughts kept returning to how the cost of storing a single bit of information has changed over my lifetime and how that has impacted almost every aspect of contemporary life.

My introduction to computing and digital storage of information would seem stone-age to anyone born after the advent of personal computers. Beginning my college education in the mid-1960s meant learning to program a central mainframe computer by inputting the data on punched cards, accessing a limited amount of data storage on the mainframe (typically a few kilobytes), and getting the output on a few pieces of paper spit out by a "teletype" printer. I recall being chastised by the instructor in my first programming class when one of my early programs included a common error–a calculation to divide by zero–resulting in the teletype printing pages of gibberish until the entire paper allotment for the class was used up. Fortunately, the availability of increasingly powerful computers and versatile computer programming languages improved rapidly as I moved through my education, protecting students and computing resources from such simple errors.

The cost of computing and storing information digitally has fallen steadily over the last 50 years, following the now well-known prediction of Gordon

© Springer Nature Switzerland AG 2020
H. Frederick. Dylla, *Scientific Journeys*,
https://doi.org/10.1007/978-3-030-55800-0_7

Moore, the co-inventor of the integrated circuit in the early 1970s. There are few predictions as prescient as Moore's—the cost of computing has fallen by a factor of two every two years since his pronouncement thanks to the ability of the semiconductor industry to double the density of the transistor switches on integrated circuits. The cost of the electronic circuits that store digital information has fallen even faster than Moore's Law.

Over the last two decades, the big "tech" companies like Microsoft, Apple, and Google have built huge and redundant data farms with the goal of storage and easy access to digitally stored information. We have all become overly dependent on search engines to find very basic information: directions, a good place to eat, answers to the Sunday morning crossword puzzle, and so forth. The instant information is satisfying! But what happens when you attempt to "google" a serious scientific topic and find hundreds of potential and often dubious leads? One of a scientific publisher's most important tasks is to more effectively filter this ever-growing digital inventory.

The prospects for enhancing discovery are enabled by the fact that almost all new scientific publications are committed to digital memory in more than one place: the author's institution, the publisher's platform, and one of several digital archiving services that have been established by the library and publishing communities. In addition, significant effort has been expended in digitizing the back file (i.e., the stored paper copies) of the pre-digital era. A modest-sized single hard drive can contain the entire database [1] of research publications (70 million articles by 2019). This task requires about 5 terabytes—think of a 5000 gigabyte thumb drive. The continuously decreasing cost of digital memory and computer processing has given us the ability to carry around a complete library in our hand.

Fig. 7.1 The original Colossus computer at Bletchley Park, 1943. *Credit* U.K. Government

Five terabytes at our command—that was the first thread of my cross connect on digital memory during this particular tour of the U.K. Two other historical events heightened my fascination with the power of cheap bits and bytes. I visited Bletchley Park, just outside the city of Oxford. Bletchley Park was a sleepy old English estate in the late 1930s when the English government bought the manor house and surrounding land as a home for its code-breaking activities.

Most of its fascinating history surrounded the successful cracking of the German Enigma codes during World War II [2]. Bletchley Park was also the home of the first digital computer, the first of 10 Colossus computers [3], built to automate the code-breaking process using a stored program algorithm developed by mathematician Alan Turing. Unfortunately, the original Colossus did not survive, but a faithful, working replica of the computer was built by dedicated amateurs who managed to find nearly all of the 70-year-old parts. This machine, with only 20 kilobytes of punched-paper-tape memory, saved thousands of lives in the last two years of the war.

Fig. 7.2 A docent from Bletchley Park in front of the rebuilt Colossus. *Credit* Linda B. Dylla

Alan Turing's genius followed me across the Atlantic as I read George Dyson's account [4] of Turing's legacy with John von Neumann at Princeton's Institute of Advanced Study. At Princeton, von Neumann constructed one of the first serious digital computers in the United States in 1953. This was a fateful year—with the first H-bomb blast at the Bikini Atoll and the publication of Watson and Crick's paper in Nature revealing that they had cracked the genetic code. Von Neumann's machine was important for both ventures: it was funded by the U.S. military to perform the laborious calculations simulating thermonuclear blasts, and it was also used by a Princeton biologist to simulate the evolution of life by the coding of proteins on DNA molecules. Dyson noted the strange juxtaposition of a single machine that was used to produce something that could both protect and destroy life. This machine had only 5 kilobytes of digital memory (using the persistence of phosphors on cathode ray tubes), and this single cache of memory represented a staggering 20% of the world's inventory of "fast" random access memory at the time. Just 20 kilobytes in Bletchley Park hastened the end of World War II; 5 kilobytes in Princeton helped keep the Cold War cold and launched a means of making biology and medicine quantitative.

What will scholars' unfettered access to 5 terabytes (and more) of stored information unleash? The discussions among the science community for completely open scientific data bases are attempting to answer this question and make it the norm of scientific practice.

References

1. This database is maintained by Crossref, a nonprofit organization established in 2000 by a group of scholarly publishers to assign "Digital Object Identifiers" to scholarly articles and related information so that such information can be easily found and interlinked; see http://www.crossref.org
2. F.H. Hinsley and Alan Stripp, "*Codebreakers: The inside story of Bletchley Park*", Oxford University Press (2017)
3. Allen W.M. Combs, "*The making of Colossus*", *IEEE Annals of the History of Computing*, **5**(3),253 (1983), https://doi.org/10.1109/mahc.1983.10085
4. George Dyson, "*Turing's Cathedral*", Pantheon Books (2012)

8

The Electronic Water Cooler

Early in Marissa Mayer's tenure as CEO of the troubled web search firm Yahoo, she issued a corporate memorandum [1] calling all work-at-home employees back to the office. Her call for "all hands-on-deck" in corporate headquarters unleashed an international debate about the telecommuting business strategy. Are group creativity and innovation better inspired by face-to-face interactions than by modern electronic connections?

Twenty years after the web moved from interconnecting high-energy physics labs to being an essential tool for commerce, high-bandwidth connectivity has certainly made it easier for employees to carry out many tasks that once required their presence in the workspace. Such remote activity has obvious advantages to child-rearing parents and commuters in congested areas. But do the advantages of this relatively new-found connectivity compensate for the loss of physical interaction in the workplace?

It should be noted that, even without telecommuting, electronic communications often suppress in-person communications because of their ease. I often urged colleagues to forgo email and walk down the hall to have a face-to-face conversation.

In responding to a thoughtful March 2, 2013, editorial published in *The New York Times* on this controversy, a letter to the editor was published by Norman Axelrod, a former Bell Labs employee, who touted his institution's iconic reputation as a hotbed of innovation—in part because of its working environment at both its Murray Hill and Holmdel, New Jersey, locations.

© Springer Nature Switzerland AG 2020
H. Frederick. Dylla, *Scientific Journeys*,
https://doi.org/10.1007/978-3-030-55800-0_8

Fig. 8.1 Bell Laboratories Headquarters in Murray Hill, New Jersey. *Credit* Lucent, 2007

The environment fostered frequent staff encounters in hallways, resource centers such as libraries, and especially in the lunchroom. When you talk to former Bell Labs employees or read the superb Bell Labs history authored by Jon Gertner [2], aptly called *The Idea Factory*, management considered real estate to be a major part of the grand design in creating a culture for personal interaction. Bell Labs' unmatched creativity also stemmed from hiring a broad array of scientists, engineers, and technicians that spanned the whole range of skills needed to develop communication technologies—a practice that became a tradition for most of the twentieth century. Moreover, the AT&T-managed monopoly with the U.S. government allowed for stable, long-term funding for Bell Labs until the court-ordered breakup of the Bell System in 1984. The Bell Labs real estate was designed to encourage and enable the interdisciplinary staff to mix both formally for the task at hand, and informally, to take advantage of a serendipitous meeting of minds.

I have had the pleasure of knowing and working with many Bell Labs colleagues over my nearly five-decade scientific career and have come to admire and envy what they experienced. I have also seen where similar cross-connections of creative people have encouraged innovative behavior. I worked for two modest-sized DOE national labs and each required a highly interdisciplinary staff. The communal lunchrooms at these two labs gave birth to more good ideas than the sum total of motivational courses to which we subjected our staff. I had the pleasure of working for the founding director of Jefferson Lab, Hermann Grunder, who stacked every lunch table with a pencil and notepad to make sure a good thought didn't lose its fidelity on a napkin. (For more on Hermann Grunder, see Chap. 25.)

One of my jobs at Jefferson Lab was fostering collaborations between the laboratory and neighboring research universities. I quickly became aware of the geographical disadvantages of modern universities, where academic departments are often enshrined in separate buildings. As I made my campus visits, I encountered two independent groups at one university doing laser-induced chemistry studies; they were separated by a street and two departmental bureaucracies. Had they talked to each other, both groups *could* have strengthened their efforts. They could have boosted their power collectively—but they didn't. At a second campus, I found a trio of scientists all working on nanocrystalline diamonds—one an experimentalist, one a device builder, and one a modeler—but none of the three had ever talked to the others about collaborating and combining their obvious strengths.

My personal experience in the sciences and engineering compels a strong bias for staff co-location—not only for the obvious tasks of designing, building, and testing machines from small instruments to gargantuan particle accelerators, but also for the day-to-day chance collaboration that creates a serendipitous solution to a shared problem. I don't see this being replaced by a virtual presence on a handheld device or laptop screen. Now, I might change my mind when my laser buddies usher in a full 3D holographic presence—but how will we share the same cup of caffeinated conversation starter?

References

1. Yahoo Orders Home Workers Back to the Office, *New York Times*, February 26, 2013, https://www.nytimes.com/2013/02/26/technology/yahoo-orders-home-workers-back-to-the-office.html
2. Jon Gertner, "*The Idea Factory: Bell Labs and Great Age of American Innovation*," Penguin Press, NY (2012)

9

Lessons from Steve Jobs

In April 1976, two young men introduced the Apple I personal computer to a group of hobbyists. One needed only to plug in a keyboard and TV monitor to make this computer work. The two men were Steve Jobs and Steve Wozniak. While Steve Wozniak designed and built this first fully self-contained personal computer [1], Steve Jobs handled sales and marketing. The product, retailing for $666.66 because Wozniak "liked repeating digits," launched what would become the most valuable company in the world.

With Steve Jobs' untimely death from cancer in October 2012 and the subsequent publication of Walter Isaacson's acclaimed biography of Jobs [2], much has been written and spoken about Job's career and management style. A master biographer, Isaacson is well known for his accounts of the lives of Albert Einstein and Benjamin Franklin. Jobs' biography tells the business mogul's fascinating story and includes information from interviews with essential personalities in the evolution of the personal computer and hand-held digital devices that allow the use of a wide range of media from almost anywhere. The preface to the book speaks of how in 2004 Jobs actively sought out Isaacson to write his biography, a classic example of Jobs' chutzpah. At the time Isaacson politely declined, noting that he would normally write biographies about people who have either departed this earth or are nearing the end of their life and career. Then Jobs' wife, Laurene Powell, contacted Isaacson in early 2009 to say that if Isaacson was going to write the biography with Steve's input, he had better get started.

© Springer Nature Switzerland AG 2020
H. Frederick. Dylla, *Scientific Journeys*,
https://doi.org/10.1007/978-3-030-55800-0_9

Fig. 9.1 Walter Isaacson. *Credit* Courtesy of Aspen Institute

Having considerable management and business skills (experience as managing editor of *Time* magazine and chairman of CNN), Isaacson is a keen and experienced observer of the unique business skills that Jobs applied during his career. A superb example of such skills manifested itself during Jobs' second tour of duty as Apple's CEO, when he rescued the company from bankruptcy, turning it into one of the highest-performing companies in the world. Isaacson summarized what he saw as Jobs' most important leadership skills in a subsequent article [3] for the *Harvard Business Review* that was published shortly after his Jobs biography. I believe that many businesses and organizations alike can use these lessons to advance their missions. Here are some of those lessons:

Focus. When Jobs returned to Apple in 1997, he directed senior Apple managers and engineering staff to focus on a few perfect products that fit a two-by-two matrix, with columns labeled "consumer" and "professional" and rows labeled "desktop" and "portable." The new Apple product line quickly fell in place. Similar focus was applied to the concept of Apple Stores— a venture Jobs successfully pushed over the objections of Apple's Board of Directors. Apple Store customers encounter the aesthetic minimalism of Apple products (MacBooks, iPhones, iPads, etc.) and appropriately trained staff—quite a different experience from "big box" electronic stores. Apple Stores are known to be the highest grossing retail outlets per unit of floor area.

Simplify. Prior to the Apple I, Jobs worked for the video game company Atari. The instructions for Atari's "Star Trek" game were simple: (1) insert 25 cents; (2) destroy Klingons. All instructions should be straightforward and intuitive, Jobs believed. At Jobs' direction, most Apple products have no need for an instruction manual.

Put products before profits/push for perfection. A product can be over-engineered, and there are points of diminishing returns. As a counterpoint, Isaacson tells the story of Jobs' obsession with every feature of the IPhone until he felt that it was perfect. To what result? The product took over more than a third of the cell phone market in a matter of months. This product-development path became the Apple standard, continually yielding substantial returns on investment. There are lessons of tolerance here for impatient boards and stockholders.

Fig. 9.2   Steve Jobs with iPhone 4. *Credit* Photograph by Matthew Yohe Wikipedia Commons, CC-BY-SA-3.0

Combine the humanities with the sciences. Although Jobs was neither a scientist nor an engineer, he was absolutely brilliant in tying together technologies that would enable consumers to write, listen to music, talk to friends, and watch a video presentation with ease. One of the few courses that Jobs took during his one-year study at Reed College was typography. His love for beautiful typeface drove Apple computer technology to include composition with attractive type. His love of music presaged a similar innovation with the launch of the iPod, and his coupling of the iPod with the iTunes music delivery service effectually saved the recorded music industry from pirated digital copies and again boosted Apple's revenue.

Isaacson's article contains more key business lessons that could be widely applied to other businesses. For example, trade publishers learned from Jobs' take-it-or-leave-it arrangements with publishing houses for delivering content on the iPad. If you don't have time to read Isaacson's superb biography, take a few minutes to read the Harvard Business Review article with its leadership lessons, some of which can be helpful when applied to pursuits other than business, such as, for example, the communication of science.

References

1. If you are lucky enough to still own one of the 200 Apple I's that were produced by Wozniak and Jobs, you'll be pleased to know that one unit was sold by Christie's auction house in November 2010 for $178,000
2. Walter Isaacson, "*Steve Jobs*", Simon and Schuster, (2011)
3. Walter Isaacson, "*The real lessons of Steve Jobs*", Harvard Business Review, April 2012;. https://hbr.org/2012/04/the-real-leadership-lessons-of-steve-jobs

10

What We Hear from Bose

On July 12, 2013, the world lost Amar G. Bose, but we will be hearing his contributions for many years to come. Professor Bose had a remarkable 45-year teaching career at the Department of Electrical Engineering and Computer Science at MIT, where he undoubtedly influenced the careers of hundreds of students who took his courses or admired his contributions to acoustics, electronics, and business. Beyond the MIT community, Bose is a familiar and admired brand name, reflecting the name of the company he founded and led through its entire history. The Bose Corporation improved the way we hear music from the solitude of an individual headphone shutting out background sounds, to mimicking near-concert-hall acoustics in your home, to a crisp broadcast at the town bandstand.

As a student at MIT in the mid-1960s Professor Bose was analyzing how we hear and appreciate music in the concert hall. He found that our ear and auditory senses process a complex mixture of sound coming directly from the source and reflecting from nearby surfaces.

© Springer Nature Switzerland AG 2020
H. Frederick. Dylla, *Scientific Journeys*,
https://doi.org/10.1007/978-3-030-55800-0_10

Fig. 10.1 Amar G. Bose, 1929–2013. *Credit* Courtesy of Bose Corporation

From this discovery came the first Bose product—the 901 speaker system that used an array of small and relatively inexpensive speakers, arranged to both beam sound directly at the listener and reflect sound from the walls of the listening environment. The Bose Corporation grew to a billion-dollar class enterprise that has sustained a globally trusted brand for sound systems of all sizes. Indeed, they are expensive, but Bose products lead the market in performance and quality. I still own and use every Bose product I ever bought.

In the interest of full disclosure, I confess that I didn't exactly buy my first Bose product. As an undergraduate at MIT, I started a modest science project in my sophomore year which, three years later, turned into my Master's thesis. The project involved very sensitive acoustic measurements that had to be made in a special room surrounded on all sides by meter-long sound-deadening foam cones. Such "anechoic chambers" are needed for accurate measurements of microphone and speaker performance. Given the need for this special facility by Bose and the other well-known MIT acoustics researchers, my use of the chamber was delegated to the proverbial grave-yard shift. Despite being largely isolated during these overnight hours, I got to know a number of MIT students who produced their own sketches of the famous 901 speakers with the intent to construct it. With the help of the local electronics parts shops and the MIT woodshop, a few enterprising students succeeded in building pretty good copies. I was one of those students. (I imagine that the best of these copycats were hired by Bose for the real thing.)

According to Bose, his company is able to maintain its strong market position by continuously investing in new product-related research and development. He believed that his company's level of R&D spending would not be possible if his corporation were publicly held, as such expenditures are subject to the whim of the daily stock market review for short-term returns. One has to ask, is this an important lesson for corporate innovation? Certainly, not having to answer immediately to market pressures gives innovators with their own financial resources more freedom to invest in R&D as they see fit. On the other hand, the markets can substantially reward a company for its investments in product innovation, if subsequent earnings are real, positive, and sustained. Apple, in its second phase of leadership by Steve Jobs, is a well-known example. A common feature of both companies is their visionary leadership in product innovation and corporate management alike.

Amar Bose is remembered for his visionary leadership as a scientist, teacher, inventor, and businessman. He touched many through his life's work—his students, the research and business communities, and all of us who enjoy listening to music in the way that it's meant to be heard.

11

The French Connection

Thanks to the telecommunications companies who strung the world with high-capacity fiber-optic cable over the last two decades, anyone with a mobile phone can communicate nearly everywhere in the world with voice and data at a negligible cost. Those born before the "millennial" generation can remember when the situation was quite different, and their parents can remember when a long-distance phone call—across the continent or across the world—was an expensive proposition. I was reminded how dramatically the accessibility of communications has changed in my lifetime during a visit to a small house-sized structure on Cape Cod. In the town of Orleans, MA, the French government built an end station for French transatlantic telegraph cables in the 1870s. This historical structure has since been turned into a museum [1] and preserved for decades by a group of local citizens.

The story of the first transatlantic cables is generally well known because success took several tenacious attempts [2]. American businessman Cyrus Field and the Atlantic Telegraph Company constructed the first cable from Newfoundland to Ireland, and the heralded first official transmission came in 1858 from Queen Victoria of England to U.S. President Buchanan. Before then, whether from head of state or citizen, all messages across the ocean had to be sent by boat—a minimum of 10 days. The telegraph took this down to 1.5 h, when working optimally. The expense of this new infrastructure was $20 million; to put this into perspective, the United States purchased Alaska the following year for $7.2 million. And to transmit one word on this cable cost $10, more than a week's earnings for a typical laborer [3]. Think about that the next time you study your phone bill.

© Springer Nature Switzerland AG 2020
H. Frederick. Dylla, *Scientific Journeys*,
https://doi.org/10.1007/978-3-030-55800-0_11

Communication was a greater commitment in earlier times. Consider the care that senders took to compose these important and infrequent messages and compare that with the hazards of today's instantly composed email or tweet.

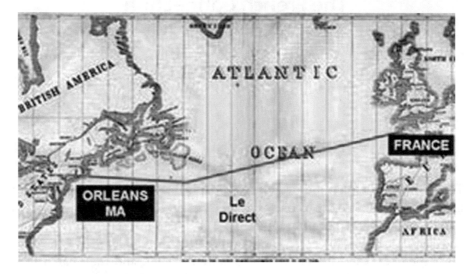

Fig. 11.1 The French connection. *Credit* Courtesy of the French Cable Station Museum, Orleans, MA

The monopoly of transatlantic communications was held by the US-British enterprise, and this caused a problem for the French. For national pride, let alone sensitive business and diplomatic communications, the French needed their own cable. Over the period 1869–1879, the French laid cables from the coastal town of Brest, to the small French-owned island of St. Pierre-Miquelon off the coast of Newfoundland. In 1890, they routed a shorter cable down to the end station on Cape Cod for distribution across the burgeoning U.S. telegraph network; they were finally able to send key messages, privately, to the banks in New York and embassies in Washington.

It's humbling to consider the technology required to send messages over those 3000-mile cables. The dots and dashes of Morse code were tapped out with telegraph keys that interrupted a battery voltage. There were no such things as electronic amplifiers. On landlines, weak signals could be interrupted and retransmitted with operators manning repeater stations. No such technology could be used on cable sunk to the bottom of the Atlantic Ocean. The solution to this problem could aptly be considered the first optoelectronics endeavor of all time, the progenitor of the semiconductor lasers that today transmit our messages along fiber-optic cables that traverse continents and oceans. The very weak electrical impulses sent via this French cable were transformed to deflections of light with a device called a galvanometer. The operator could record the duration or direction of the deflection as a Morse dot or dash. Can you imagine how tiring or prone to error such a task was? This led scientists to develop a recording device whereby the galvanometer dragged a minute glass fiber pen across a roll of paper. Now the message was archived and could be checked for errors.

Fig. 11.2 The French Cable Station Museum, Orleans, MA. *Credit* Courtesy of the Museum

This sounds primitive compared to the technology that enables us to message instantly today. It is remarkable to consider that this French cable station stayed in operation from 1890 until 1959, when the function was made obsolete by better technologies, including AT&T's transatlantic telephone cables. The first telephone cable in the series used 51 electronic repeaters (based on miniature vacuum tubes) and offered 36 channels for better, faster messaging across the ocean. Our present transatlantic calls and web traffic are now routed through fiber-optic cables. The first fiber-optic cable, the TAT-8, was commissioned in 1988; since then many more cables have been laid. The fiber optic cables that cross the Atlantic today deliver 10,000 times more information in one second than the French cables tied into Orleans delivered in nearly a century of operation.

References

1. Warren Darling, "*The French Cable Station Museum*", Lower Cape Publishing Orleans, MA (1988) http://www.frenchcablestationmuseum.org/index.htm
2. Tom Standage, "*The Victorian Internet: The Remarkable Story of the Telegraph and the Nineteenth Century's On-Line Pioneers*", Bloomsbury Publishing, London (1998)
3. Andrew Odlyzko, "*The history of communications and its implications for the Internet*", Computer Networks **36**, 493–517 (2001) http://papers.ssrn.com/sol3/papers.cfm?abstract_id=235283

12

Mr. Fresnel's Gift to the World

Many of us try to escape the heat of the summer months in the Northern Hemisphere by catching the cool breezes of the shorelines. If you find yourself at the beach and want to enjoy something else that's cool (for science, history, and even art), I recommend that you seek out a lighthouse. Most of these structures peppering the coastline are historic properties, once essential tools of commerce long ago, now replaced by automated hazard markers, radio beacons, and GPS receivers.

Fig. 12.1 Lighthouse at Dover Castle in Kent, England, built by the Romans soon after they arrived in Britain, circa 50–138 AD. *Credit* Chris McKenna (Thryduulf), Wikipedia Commons, CC-BY-SA-4.0

© Springer Nature Switzerland AG 2020
H. Frederick. Dylla, *Scientific Journeys*,
https://doi.org/10.1007/978-3-030-55800-0_12

Lighthouses have been recorded in history since the time of the fabled structures built by the Greeks in the third century BC in Alexandria and Rhodes. For nearly 2000 years, lighthouses were simply towers or piles of rocks where a fire or torch was kept lit to guard a dangerous shoreline. By the early eighteenth century, both sides of the English Channel and much of the American coastline were dotted with grand lighthouses marking dangerous conditions. However, the persistent problem throughout most of this bi-millennium was the all-too-short warning time. In many cases, wary sailors were unable to spot the dim warning beacon in time and found themselves in the midst of the very hazard that they were being cautioned against. By the late eighteenth century, burning piles of wood had been replaced with a whale-oil lamp and simple, parabolic mirror, but the observable range in fair weather was still only a few miles. These beacons may have lured more sailors to their deaths than they actually protected. The situation was changed radically by Augustin Fresnel, a young French engineer, who invented a compound lens that efficiently focused a large fraction of light from a single source into a concentrated beam. This famous lens bears his name.

Fig. 12.2 First-order lighthouse Fresnel lens, on display at the Point Arena Lighthouse Museum, Mendocino County, CA. *Credit* Frank Schulenburg, Wikipedia Commons, CC-BY-SA-3.0

Fresnel had that unique combination of a knowledge of physics and the ability to perform precise calculations as well as a tenacious ability to convince French glass manufacturers to develop improved methods of making accurately shaped and sufficiently pure glass lenses and prism elements. Armed with these new lenses, his first public demonstrations were successful. When his first lens was installed on the French coast at Cordouan, the resulting beacon was 38 times brighter and could be seen by sailors on the top rigging of their ships 33 nautical miles away. As an additional benefit, Fresnel calculated that his light consumed half as much oil as one of the simple reflector lights on the neighboring English coast. A typical Franco-British trade war followed over the next few decades as the sailors, their underwriters, and new industries clamored to build and install the new lenses on both coastlines. In France the number of shipwrecks each year dropped by two-thirds.

The lighthouse situation in the new nation of the United States provides an almost comical example of Yankee frugality. By the time the U.S. secured its independence in 1776, there were a dozen lighthouses on eastern shores. The new U.S. Congress did not let much time pass before usurping control of these structures (and imposing shipping revenue taxes) under the Lighthouse Act of August 7, 1789. During the first three administrations, the commissioner of revenue controlled the service. Unfortunately, when the specific office of the U.S. Lighthouse Establishment was created by President James Monroe in 1820, control was assigned to an Administration accountant, the "Fifth Auditor of the U.S."—Stephen Pleasanton. Mr. Pleasanton did not live up to his surname; for the next 30 years his mantra was to save funds, even if it came at the expense of commerce and poor infrastructure. He had no scientific or engineering knowledge and had never been to sea to experience at first-hand the purpose of these maritime sentinels. Pleasanton resisted expenditure for a Fresnel lens; he was proud that he spent less money than the U.S. Congress appropriated for the task of operating a larger number of lighthouses than Britain—and at a fraction of Britain's cost. Never mind that these lights couldn't be seen. The maritime community constantly complained of their inferiority compared to what they experienced on the other side of the Atlantic.

It took a specific act of Congress in 1838 and the involvement of a growing technical class—engineers from the Navy and West Point and physicists from Princeton and Harvard—to circumvent Pleasanton's control. The first U.S. beacons to use Fresnel lenses were lit off Nantucket and in the Delaware Bay at Brandywine in 1850; the lenses garnered immediate praise for their sight range of 40 miles.

A comparison of the Fresnel lamp at Brandywine with the simpler reflector lights of Cape Henlopen to the south showed that the Brandywine light was nearly six times brighter and used nearly four times less oil than the Henlopen light. The latter argument swayed Congress to establish a new Lighthouse Board in 1852. The economic argument was aided by the frightening experience of several Congressmen on a steamer from Washington to New York harbor—the ship's captain could not make out the traditional harbor lights off Sandy Hook, NJ. The new board was aided by physicist Joseph Henry, who was asked to chair the committee on experimental modeling.

Clearly Fresnel's applied physics trumped politics on both sides of the Atlantic. If you would like to learn more about this fascinating man and his legacy, I can recommend historian Theresa Levitt's book "A Short Bright Flash," [1]. You can draw your own conclusions about whether shining light on Congressional actions has improved since Fresnel's lamp was lit on the Delaware Bay.

Reference

1. Theresa H. Levitt, *A Short Bright Flash, Augustin Fresnel and the Birth of the Modern Lighthouse*, W.W. Norton & Co., NY (2013)

13

Lighting the Way for Innovation

The world lost a remarkable scientist on January 27, 2015 when Charles Townes, who is most usually associated with the invention of the laser, passed away just six months shy of his 100th birthday [1].

Fig. 13.1 Charles H. Townes, 1915–2015. Courtesy of AIP Emilio Segrè Visual Archives, *Physics Today* Collection

Townes' scientific journey that led to the concept of the laser began with his work at Columbia University on molecular spectroscopy using microwaves. The first intense microwave sources were invented during World War II and applied to early radar systems. In the postwar period, microwave technology

© Springer Nature Switzerland AG 2020
H. Frederick. Dylla, *Scientific Journeys,*
https://doi.org/10.1007/978-3-030-55800-0_13

was continually advancing for communication systems that took advantage of the higher information-carrying ability of these higher-frequency radio waves. That evolution continues today as our burgeoning cell-phone system takes advantage of gigahertz-range transmission over the short distances between handsets and cell towers.

Fig. 13.2 Charles Townes and his colleague Arthur Schawlow. Courtesy of the AIP Emilio Segrè Visual Archives

In his oral histories archived in the American Institute of Physics' Niels Bohr Library and Archives [2], Townes recounts the ah-ha" moment that occurred as he was sitting on a Washington, DC, park bench in 1951, pondering the problem of how to make more intense microwave radiation than could be obtained with the usual source of any form of electromagnetic radiation—a body heated to a certain temperature which emits radiation according to a well-defined spectral curve called the black body curve. By using radiation that is emitted by atoms or molecules as they are first excited to higher energy states and then forced to emit in unison, the limits of this black body curve could be greatly exceeded, and the emitted radiation could be tuned to specific wavelengths. This idea was brought to practice in the first MASER—Townes' acronym for Microwave Amplification by Stimulated Emission of Radiation. Townes and his brother-in-law, fellow Nobel laureate Arthur Schawlow, then coauthored a seminal 1958 *Physical Review* paper, "Infrared and Optical Masers" [3]. The paper predicted that the maser effect could be extended to infrared and visible light wavelengths. Thus, the concept of the laser was born.

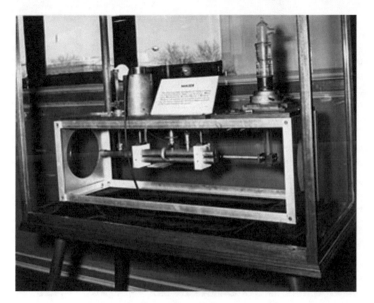

Fig. 13.3 Original masers of Dr. Charles H. Townes on display at the Franklin Institute. *Credit* J. J. Barton, The Franklin Institute, courtesy of AIP's Emilio Segrè Visual Archives

The first experimental demonstration of the laser by Theodore Maiman at Hughes Aircraft in 1960 involved a device with a flash-lamp-excited ruby crystal that produced bright red laser light at 634.8 nm. The world celebrated the 50th anniversary of the laser in 2010 [4] with all the ceremony appropriate for what is clearly one of the most important inventions of the last century. The laser is now ubiquitous: responsible for transmitting nearly all the world's information through optical fiber networks, taking in information with scanners, delivering that information with laser printers, fabricating microelectronics, cutting and welding metals for autos and ship hulls, restoring our sight, and treating certain cancers. Those are just a few of the many applications that resulted from those contemplations on a park bench.

I can trace both ends of my scientific career to Townes' invention. As a young boy I was fascinated with the demonstration of the first ruby laser and spent three years building three different working models that convinced me I wanted to be a physicist. (I recounted my adventure in an article published in *Physics Today* [5] during the laser's anniversary year, and there is an expanded version in Chap. 16).

Fig. 13.4 President Ronald Reagan awards Charles Townes with the National Medal of Science in May 1983. *Credit* White House photographer Jack Kightlinger, courtesy of AIP's Emilio Segrè Visual Archives

At the other end of my career, I spent 15 years with my colleagues at the Thomas Jefferson National Accelerator Facility in Newport News, VA, building a "free electron laser" that still holds the world's record for delivering the most power over a broadly tunable range from the infrared through the ultraviolet [6].

The American Institute of Physics honored Townes with the 2000 AIP Science Writing Award for his book, "*How the Laser Happened: Adventures of a Scientist* " [7]. I was fortunate enough to attend the award ceremony and when I asked Townes to sign my copy of his book, I was able to thank him for the wonderful invention that he gave the world—an innovation that has now become his legacy. I join thousands of fellow scientists and engineers who consider Townes to be a guiding light of our careers.

References

1. Charles H. Townes obituary, *New York Times*, January 29, 2015; http://www.nytimes. com/2015/01/29/us/charles-h-townes-physicist-who-helped-develop-lasers-dies-at-99.html?_r=0

2. AIP Niels Bohr Library, Oral Histories: a). Interview with Charles Townes by William V. Smith, June 18 and 20, 1979; https://www.aip.org/history-programs/niels-bohr-library/oral-histories/4653 b). Interview with Charles Townes by Joan Bromberg, January 28 and 31, 1984; https://www.aip.org/history-programs/niels-bohr-library/oral-histories/4917-1 c). Interview with Charles Townes by Finn Aaserud, May 20 and 21, 1987; https://www.aip.org/history-programs/niels-bohr-library/oral-histories/4918
3. A. L. Schawlow and C. H. Townes, Phys. Rev. 112, 1940 (1958); https://doi.org/10.1103/PhysRev.112.1940
4. "LaserFest" the international commemoration of the 50th anniversary of the demonstration of the first laser; http://www.laserfest.org
5. H. Frederick Dylla, "*My First Laser*", *Physics Today*, August 12, 2010, https://physicstoday.scitation.org/do/10.1063/PT.4.0068/full/
6. H. Frederick Dylla and Steven T. Corneliussen, "*Free-Electron Lasers Come of Age*", Photonics Spectra, August 2005; https://www.photonics.com/Articles/Free-Electron_Lasers_Come_of_Age/p5/v19/i120/a22472
7. C.H. Townes, "*How the Laser Happened: Adventures of a Scientist*", Oxford University Press, Oxford, UK (1999)

14

China's Science Ambassador

In 1990, I accepted a position with a new U.S. Department of Energy laboratory, now called Jefferson Lab, in Newport News, VA. Within nine months of my arrival, I was leading a technical group responsible for producing nearly a kilometer of superconducting acceleration modules. These devices became the engine of a four billion electron volt accelerator whose electron beam is used to probe the structure of the atom's nucleus. In 1996, the facility became operational as a research center used by more than one thousand nuclear physicists worldwide. That same accelerator technology, now integral to many large international scientific user facilities, is used to probe the mysteries of ordinary materials and subatomic matter. For the past 70 years, such facilities have promoted partnerships and collaborations among scientists.

While immersed in the construction project at Jefferson Lab, I met Professor Chen Jia'er. He was the head of Peking University's Institute of High Energy Physics and would periodically visit Jefferson Lab to establish collaborations between Chinese and American physicists. At the time, I had no idea of Professor Chen's essential role in rebuilding the Chinese scientific establishment after the Cultural Revolution ended in 1976. Members of Chinese academia had to begin anew in all aspects of their careers. As China began to rebuild its economy and relax travel restrictions, pioneers like Chen began developing essential ties with the international scientific community.

Fast forward 20 years, and Chen became a revered senior scientific statesman having served as head of the National Natural Science Foundation of China (NNSFC) and as President of China's most prestigious academic institution, Peking University. Over those two decades, China has become the world's second largest economy, established a countrywide network of

© Springer Nature Switzerland AG 2020
H. Frederick. Dylla, *Scientific Journeys*,
https://doi.org/10.1007/978-3-030-55800-0_14

research universities and currently educates more than 30 million undergraduates. One important measure of the national scientific productivity is the number of scientific publications authored by its scientific community. In 2007, I became head of the American Institute of Physics (AIP), a scientific organization whose primary activity is publishing journals in physics and allied fields. I witnessed the rapid rise of Chinese scientific productivity while monitoring the authorship of AIP's journal content. By 2009, Chinese authors submitted more articles to AIP journals than authors from any other nation including the United States. This growth in scientific publications has continued over the last decade in proportion to the continued strong investments that the Chinese nation makes in science.

I had a first-hand look at the Chinese scientific enterprise when I helped AIP open its first international office in Beijing in 2010. We invited nearly 100 guests from Chinese academia, business, and government to help AIP celebrate the official opening of our office. Framed between the celebration's announcements and entertainment from traditional and modern Chinese operatic troupes was the celebration's highlight: Professor Chen's brief comments from the podium. He noted that collaboration among scientists from the two countries has had great influence on the growth of positive U.S.—China relations in general, and he thanked AIP for building bridges with the international reach of our communications and publications. His remarks to the audience, which included many of his colleagues and former students, received great attention and respect.

On our last evening in China, AIP hosted a small private dinner with Chen. We reminisced quietly about how China has transformed itself on the world stage, particularly in the last decade. In the neighboring room, a group was having a much noisier dinner—clearly a celebration of sorts. Suddenly, the room quieted and one by one the neighboring group paraded to our table to pay their respects to Professor Chen. The group included many of his former students, now professors themselves. They were celebrating a young physicist's first award from the NNSFC—a program designed and established by Chen. We could not hope for a better ambassador for China's science than Chen Jia'er, who pledged to help his American guests and other scientific publishers promote the best of China's physics to the rest of the world.

Fig. 14.1 Professor Chen Jia'er. *Credit* American Institute of Physics

15

Bringing Science Back to Vietnam

Jean Trân Thanh Vân, or simply "Van" among his friends and acquaintances, left his home country at the age of 17 to escape the Vietnam War. He went to France to pursue a career in physics, which was his dream. His accomplishments in theoretical particle physics would suffice to categorize his career as successful, yet perhaps most influential has been his tireless efforts on behalf of the scientific community. Professor Van and his wife, Professor Le Kim Ngoc, broadened international participation in particle physics, astrophysics, and biology. Moreover, they were instrumental in bringing higher education, scholarship, and science back to a war-torn Vietnam. In April of 2012, it was my honor on behalf of the American Institute of Physics (AIP) to meet Jean Trân Thanh Vân and present him with the prestigious Tate Medal for international leadership in physics.

Professor Van obtained undergraduate and graduate degrees in mathematics and physics, establishing a career in particle physics in the mid-1960s. At this time, this field of study was bursting with the first key experimental results on the underlying (quark) structure of the proton and the theoretical foundations of the now "standard model" of particle physics. Shortly after getting his doctoral degree, Professor Van began organizing workshops in Paris ("Rencontres") between groups of experimentalists and theoreticians from various nations to discuss frontier problems, without regard for international or political barriers. Some of the first meetings, held during the height of the Cold War, provided forums for Western and Soviet scientists to meet and interact.

© Springer Nature Switzerland AG 2020
H. Frederick. Dylla, *Scientific Journeys*,
https://doi.org/10.1007/978-3-030-55800-0_15

Fig. 15.1   Jean Trân Thanh Vân accepting the Tate Medal for international leadership in physics (2012). *Credit* American Institute of Physics

In the early 1990s, when Vietnam started to recover from decades of war and to reopen its borders to the rest of the world, Van and Kim began a three-decade campaign to bring science back to Vietnam.

The American Institute of Physics established the Tate Medal in 1959 in honor of John "Jack" Tate for his role in broadening international participation in physics in the post-World War II era. During his career, Tate served as the president of the American Physics Society (APS) and as editor of the prestigious APS journal *Physical Review*, and helped to found AIP in 1931, to rescue the scientific publication business of five scientific societies–a business that was suffering from the financial ravages of the Great Depression.

The Tate Medal is not given at regular intervals, as the award's peer-selection committee sets a high bar for honorees. After meeting and spending time with Professor Van, I am convinced that the selection committee that met in 2012 to consider potential awardees recognized that Professor Van's world-changing achievements particularly well exemplify the spirit for which this award was intended. After Professor Van received his award during the presentation ceremony at an annual meeting of the American Physical Society, I was invited to an informal dinner hosted by Van and Kim for their family and friends from the physics community. I was able to learn more about Van and his colleagues' dedication to what they view as their "last project," before they turn it over to the next generation. Many years of behind-the-scenes work with friends, colleagues, diplomats, and bureaucrats had a remarkable payoff in 2013 when the International Center for Interdisciplinary Science and Education [1] opened in Quy Nhon, a beautiful coastal town in Vietnam with two nearby universities. The Center hosts numerous international conferences every year, stressing the interdisciplinary nature of frontier science and the need for global participation.

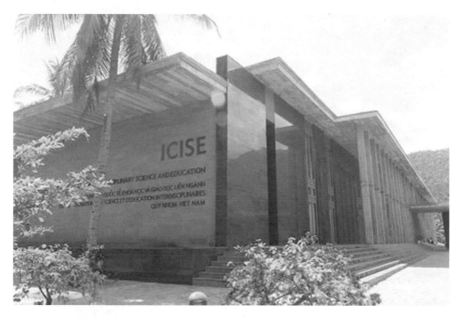

Fig. 15.2 The International Center for Interdisciplinary Science and Education (ICISE) in Quy Nhon, Vietnam, which opened in 2013. ICISE is a project of the *Rencontres du Vietnam*, chaired by Jean Trân Thanh Vân. *Credit* ICISE

I expect that the Vans and their chosen successors will continue to receive considerable support from members of the international scientific community in identifying the financial and human capital that will help sustain their ambitious creation.

Reference

1. International Center for Interdisciplinary Science and Education, Quy Nhon, Vietnam; https://www.icisequynhon.com

Part II
Mentors and Milestones

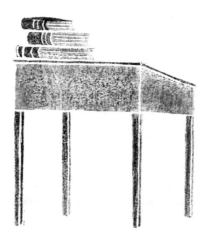

16

Exciting the Imagination: My First Laser

Within every scientist is the drive to understand how things work and to discover new ways of doing things. Yet the fervor for discovery is cultivated by strong mentors who encourage, steer, and challenge the budding scientist until (and often well beyond) he or she becomes a professional scientist. My own fascination with science was fueled by the invention of the laser and guided by scientists who took a personal interest in my curiosity—mentors who had a tremendous impact on my life and career.

When Theodore Maiman [1] introduced the ruby laser on May 16, 1960, with pulses of bright, coherent red light from his laboratory at Hughes Research, I was an 11-year old "Sputnik" kid, playing dangerously with homemade rockets and radio circuits. Although I was too young to pay much notice then, I got hooked on lasers two years later when I read an article in *Popular Science* magazine titled "*The Incredible Ruby Ray*" [2]. It thoroughly captivated me—I just had to make a laser for myself! The trouble with such a venture for a 13-year-old boy was the required equipment: a cigarette-sized ruby crystal and a high-energy flashtube far exceeded my discretionary funds. Cash from delivering papers and mowing lawns could keep a junior scientist stocked with chemicals and radio parts, but the components for a pulsed ruby laser would require a major bequest from a rich relative. I didn't have one.

An abbreviated version of this article first appeared in *Physics Today Online*, 12 August 2010, to help commemorate the 50th anniversary of the first demonstration of a laser.

© Springer Nature Switzerland AG 2020
H. Frederick. Dylla, *Scientific Journeys*,
https://doi.org/10.1007/978-3-030-55800-0_16

After I had read the article in *Popular Science*, I started reading any article I could find on lasers and spending afternoons and weekends in Philadelphia's Franklin Institute library. I needed to find someone who would lend me the required ruby crystal. By 1963, a small number of companies in the United States were making those precisely grown crystals, but the price for the required 5 cm-long specimen was well over a $1000—out of my budget. So I wrote letters to every such company that I could identify, explained my plans, and inquired whether I could borrow a ruby crystal. To my delight, a research team at RCA's engineering research facility in Camden, NJ, wrote me back and invited me for a visit. I took that to mean that they wanted to check me out before offering to help. I left the RCA laboratory with not one but two laser crystals, in addition to lots of advice on how to build my first laser. The engineers I met during my first visit to RCA stayed in touch with me during what became a four-year venture, as my lasers became increasingly sophisticated (and better working) and my ability to ask better questions matured.

I never even bothered to take a photograph of the first laser I built (model 1)—it was crude looking even to a 13-year old. These types of lasers required only three things: a cylindrical shaped ruby crystal of the right characteristics with precisely ground faces; carefully deposited silver plating on the ends of the crystal to allow light to bounce back and forth (or resonate within the crystal) and a means of producing an intense flash of white light to "pump" energy into the ruby crystal. I had a suitable crystal courtesy of RCA but no suitable flash lamp. I tried writing letters to all three companies that were manufacturing such lamps at that time, requesting a loan, but I came up empty-handed in the loan department. My meager allowance did allow me to procure a carton of pre-flashlamp era photographic flashbulbs from the local camera store. Ugly model 1 consisted of the precious ruby crystal being suspended within an empty and carefully cleaned tin can that previously contained "V-8" tomato juice. I found a tin funnel that conveniently held the ruby crystal in the center of the can and directed the desired laser light out of the other end. The outer perimeter of the V-8 can held 16 flashbulbs all pointed toward the ruby crystal. When I fired all 16 flashbulbs, I suspected the laser of having fired because I observed a bright, but tightly contained red spot on the far wall. However, my weekly budget allowed precious little experimental time—perhaps one shot a week. Clearly, I had to move on to model 2.

Having failed in my first attempt of borrowing from corporate America not only the required sophisticated flashlamp, but also the necessary power supply to fire the flashlamp, I tried my luck in the surplus electronics market. This meant trekking through junk stores with piles of largely World War II surplus electronics (this was well before e-Bay). I did find a small helical flashlamp and associated power supply that was originally built for aircraft signaling. By surrendering four weeks of my hard-earned income, the equipment was mine. Model 2 needed a more sophisticated arrangement for holding the crystal within the center of the helical flashlamp and the ability to focus as much of the flashlamp light into the crystal as possible. I was basically working to reproduce the 1960 arrangement that Maiman used for his first ruby laser demonstration. With a little help from my high school metal shop teacher, I built a much better set-up than my crude tomato can version. And now I had a laser that could be fired at will, with a push of a button, without evaporating my weekly budget. But apart from doing some crude laser tricks like burning the black ink off typewritten words on white paper and popping a blue balloon (but not a red balloon) because blue objects absorbed red light, model 2 didn't seem like a scientific tool unless I could do something with it to explore some facet of nature.

By the time model 2 came along, I was a freshman in high school studying biology. I wasn't too interested in biology because by then my laser fascination had taken over my brain and I was sure I wanted to be a physicist. I was fortunate to have had a remarkable high school biology teacher, Mr. John D. Jones, who taught me about the intrinsic value of inquiry and the fun of discovery behind any science. He urged me to use my love of physics in the study of biology. So it was back to the Franklin Institute library for me, where I turned my attention to wondering what effect laser light could have on living things. Normally, visible light doesn't carry enough energy to harm biological systems unless the light is powerful enough to cause thermal damage or burn tissues—what we typically know as sunburn. As the wavelength of light shortens into the ultraviolet and x-ray range, the light carries enough energy to damage sensitive biological molecules such as DNA. Such radiation-induced effects on DNA can cause mutations in cells, which is why precautions have to be taken with exposure to such radiation.

I did find that there was a class of experiments showing that visible light could have effects on biological systems if the absorption of light were enhanced by certain classes of light-absorbing chemicals or dyes. None of these experiments had been tried with laser light, which has both a spectral purity and the ability to focus to a tight spot that normal light sources do not have. I was clearly in over my head on this line of enquiry, given my state of knowledge of both biology and chemistry. My letter writing campaign to solicit components for my lasers had taught me that there is value in seeking expert advice. One of the world's experts on the effects of radiation on biological materials, was Professor Hermann Muller [3] from Indiana University, who had won the 1946 Nobel Prize in Physiology for his work showing that sufficient x-ray exposure could cause damage to DNA, leading to mutations in subsequent generations. I wrote to Professor Muller and asked for his advice as to whether visible laser radiation (because of its spectral purity and ability to tightly focus) might have interesting biological effects when the absorption was enhanced with a dye molecule [4]. To this day, I marvel at the letter that was sent back to me—a 15 year-old high school freshman at the time. I received a 3-page letter including background tutorials, ruminations on my query, suggestions for my experiments, and a long list of suggested reading.

With Prof. Mueller's mentoring and some more homework in the Franklin Institute library, I designed an experiment around a common dye that is available in every high school biology or chemistry lab—methylene blue. This dye is typically used to stain thin slices of biological material so that the sub-cellular parts, like the nucleus, are visible upon examination under a microscope. This dye is generally harmless at low concentrations, which meant it could be used on living specimens. And most importantly, since it was blue to the eye, its chemical structure made it strongly absorbing to red light—the wavelength of light emitted by a ruby laser.

On to year three of my formative laser adventure. I designed an experiment to test my biological hypothesis. What would happen to a simple biological system that was soaked with methylene blue and exposed systematically to ruby-red laser light. I also felt this level of sophistication cried out for an improved laser, so model 3 was born. I had finally succeeded in getting one of the laser flashlamp companies to send me a modern flashlamp. It was a reject from the production line, but still a quite acceptable version of a sleek linear flashlamp that matched the geometry of my ruby crystal. I went back to my high school metal shop to make a state-of-the-art elliptical reflector which would efficiently couple a high fraction of the flashlamp light into the laser crystal. My high school metal shop teacher didn't have a large enough piece of metal available to make such a gadget, so he begged off and I resurrected on my solicitation campaign. Fortunately, there happened to be a neighbor employed by Reynolds Aluminum Corporation and he located a surplus 15 cm cylinder of aluminum large enough for the model 3 laser cavity. Model 3 also had third-generation electronics, a big improvement over the batteries in model 1 and patched together surplus electronics in model 2. Through the entire evolution of these ever more sophisticated models, the loaned RCA crystal worked its charm. I could fire my laser at will.

Professor Muller's recommendations led me to conduct a series of experiments with my model 3 laser that involved controlled exposures of frog eggs harvested from my backyard pond and onion roots dug up from the garden. My experiments won me some minor awards in the regional science fair in Philadelphia, but, more importantly, the experience taught me how to conduct experiments. Ten years later with a freshly minted PhD in physics, and ever since, I have tried to be generous with my time whenever a young student sends me an inquiry or asks me for advice or for a loan of scientific gadgetry. I remember how the RCA engineers and Professor Muller took interest in me and how they influenced my career in a positive way. And I still have that loaned ruby crystal....

Fig. 16.1 My model 3 pulsed ruby laser (1965). *Credit* author's photo

References

1. Theodore H. Maiman, *The Laser Inventor*, Springer International Publishing, Cham, SW (2018)
2. C.P. Gilmore, *The Incredible Ruby Ray*, Popular Science **181**, **3,** 89–92 (Sept. 1962)
3. Herman J. Muller Nobel prize citation 1946; https://www.nobelprize.org/prizes/medicine/1946/summary/
4. A decade after my crude experiments with ruby laser light and light-absorbing dyes in biological systems, the process of photodynamic action was being used for basic biomedical research and the development of "photodynamic therapy" for cancer and other disease treatments. See for example: Dolmans, D.E., Fukumura, D., and Jain, R.K. *Nat. Rev. Cancer* **3** (5), 380–387 (2003)

17

The Master Teacher

Most of us can name at least one special teacher who had extraordinary influence on our lives. I was privileged with being trained and mentored by several remarkable teachers who provided guideposts for my life in science. One of the people who had unforgettable influence on my education and subsequent career as a physicist was MIT physics professor, John G. King. John was a charismatic and creative experimental physicist in addition to being an extraordinary educator. I had the benefit of being educated and coached by him during my eight years of undergraduate and graduate study at MIT from 1967 to 1975. After I left MIT as a newly minted post-doctoral student and went on to life-long career in physics, the lessons from my experiences with King remained. I try to instill his passion for learning and the value of the scientific method to all within my sphere of influence.

John G. King was born in London and educated in France, Switzerland, and the U.S. This early exposure to European cultures was evident in his view of the world. King spent his entire professional career at MIT starting out as an undergraduate in 1943 until he died in 2014 as an emeritus professor at the age 88.

John completed his undergraduate degree in physics following war service for the U.S. Army and the U.S. Navy, and at the Harvard Underwater Sound Lab. King was a student of Jerrold Zacharias who began the Molecular Beam Laboratory in the Research Laboratory of Electronics at MIT shortly after World War II. He played a significant role in this laboratory's key achievement in this era: the development of the atomic clock based on microwave

Portions of this chapter appeared in John G. King's obituary by Rainer Weiss and H. Frederick Dylla, in *Physics Today* **67**, 11, 68 (2014).
© Springer Nature Switzerland AG 2020
H. Frederick. Dylla, *Scientific Journeys*,
https://doi.org/10.1007/978-3-030-55800-0_17

spectroscopy of gaseous cesium atoms. He converted a device that was basically a laboratory experiment into a commercial and reliable instrument which became the universal time standard for the next half century.

In the early 1960s, John became the director and principal investigator of the Molecular Beam Laboratory. He transformed the research conducted there. The new research ventures applied molecular beam techniques to cosmology, low temperature physics, and biophysics. Over 100 undergraduate and 25 doctoral students obtained their degrees working on these topics during John's tenure at the laboratory. Fortunately, I was one of his students.

John's best-known experiment, still found on the first page of most electricity and magnetism textbooks, is the measurement of the charge neutrality of atoms. Even from our first exposure to science in primary school, we learned that the negative charge of the electron is equal and opposite to the positive charge of the proton. So, the two charges cancel out and the net charge of even the simplest atom, hydrogen with one electron and one proton, sums to zero. But how exact is that cancellation? John's famous experiment showed the cancellation to be extremely good—less than one part in 10^{-20} of the charge on a single electron [1]. The experiment had been prompted by a conjecture that the expansion of the Universe was due to a slight charge unbalance [2]. Another reason for performing this measurement is to establish how well we can measure such a fundamental equality as the magnitude of the charge on the proton and electron. These are such different particles–but their charge equality is a tenet of our current understanding of the sub-atomic world embodied in the Grand Unified Theory.

At about the same time as John's neutrality measurement, he conceived an imaginative experiment, again prompted by cosmological ideas, to set a hard limit on the possibility that matter, over cosmological time, begets new matter, a version of what was once called the steady state cosmology [3].

These two measurements comprise a curious class of experimental measurements that attempt to measure something as carefully as possible when we don't expect a detectable signal. They are called "null experiments" and they are often useful for confirming how well we understand a fundamental prediction of a theory. These experiments are notoriously difficult to do because there are always sources of noise that limit and mask the true-unobtainable null in any real measurement. These experimental challenges appealed to John throughout his career and they were often used to teach his students how to design and execute difficult experiments.

When I was first asked by John to join his laboratory group during my sophomore year at MIT, I was soon exposed to the customary lunchtime gathering of his graduate and undergraduate students. During one of my first meetings with the group, the conversation turned to whether John's neutrality experiment, his famous textbook result, could be improved upon. By the end of the lunch, the group had sketched an outline of a different technique that just might accomplish this task. I needed a project and was too naïve to realize what sort of challenge I was accepting. And even if I was unsuccessful, it was very quickly clear to me that I was going to learn a lot from John and his students by my attempt at a null experiment. My naiveté wore off quickly as I soon realized my chosen venture was not a casual semester-long project. Three years later, I completed a measurement that essentially equaled John's original measurement, but importantly used an entirely different scheme. I wrote my first scientific paper with him [4] as I completed my undergraduate education. Our result published in 1973 remained the established record for this null experiment, including holding the world's most sensitive measurement of electric charge for the next 40 years until a group of Italian physicists published a similar result [5] (and also pointed out a few errors in my original analysis).

One of the many things I admired about John was his desire and ability to periodically change the focus of his research efforts. He would tell his students that he thought any reasonably accomplished scientist could master a field of study in about 5 years–but wouldn't that be boring to go on studying the same thing year after year when the world is so full of unanswered questions?

During my undergraduate exposure to John's research group, his research focus was moving on from various atomic physics ventures using molecular beams to low temperature physics. A new group of graduate students was designing and building a whole series of experiments to investigate the curious properties of liquid helium at temperatures near absolute zero where helium exhibits superfluid behavior. By combining cryogenic and molecular beam techniques and observing variations in evaporation and scattering behavior, John's group invented experiments to explore the fundamental properties of superfluid helium.

By the time I was finishing my undergraduate years at MIT, John was starting another major change in research interests–to biophysics. And I wanted to be a part of this venture, so I decided to stay at MIT for my graduate years with John as my research advisor.

He began this venture by asking several accomplished scientists to join the group to explore the limits of microscopy. This first led to a series of theoretical and experimental studies to improve the resolution of electron microscopy to atomic dimensions. By applying atomic beam techniques to microscopy–particularly biological surfaces, John invented what he called a "molecular microscope," using water molecules rather than light or electrons as the illuminating projectile. His idea was to map the locations where water would evaporate or stick on small biological samples, such as cells with biologically interesting spatial resolution.

I devoted my Ph.D. research thesis project to developing the first working model of the molecular microscope. In keeping with John's healthy habit of grooming a research team, I was joined by a new crop of fellow graduate students who worked on various aspects of this potentially new microscopy technique. By the time I was ready to defend my thesis and move onto a post-doctoral fellowship at Princeton, I had produced several crude images from some model biological surfaces. At Princeton, I joined the University's Plasma Physics Laboratory as the third non-plasma physicist hired to help solve the thorny problem of minimizing the interactions in magnetic fusion devices of hot plasmas with the material surfaces of the containment vessels. John's broad training put me in good stead to tackle this problem. In addition, I was fortunate to be able to procure (and not have to build) state-of-the-art surface analysis equipment to help study these problems. Several years later, I published two more papers with John, including one in the prestigious journal *Nature* that showed micrographs of my own red blood cells imaged with desorbed water molecules [6]. That was my last direct collaboration with John, but his scientific spirit never left me. He followed my career as I moved from plasma physics to accelerator physics to laser physics to my last career stop as head of the American Institute of Physics. I was on an 8-year cycle of change—I couldn't quite manage John's 5-year reinvention cycle.

Fig. 17.1 Professor John G. King in front of the author's "molecular microscope" at MIT, circa 1972. *Credit* Massachusetts Institute of Technology

All through John's research career, he devoted considerable efforts to investigating and improving the means of teaching science. With other educators in the late 1950s and 1960s, John worked on the revitalization of high school physics, following the startling realization on the part of Zacharias that "while students had taken physics, they didn't understand anything."

When the 1957 launching of Sputnik spurred a nationwide alarm and allocation of money to improve science teaching, King became deeply involved. In cooperation with the influential Physical Science Study Committee (PSSC) [7], he produced—and acted in—eight physics movies including *Times and Clocks*, *Interference with Photons*, *Size of Atoms from an Atomic Beam Experiment*, and *Velocity of Atoms*. One of the films featured John demonstrating a principle of physics by driving his meticulously restored Bugatti automobile down an interstate highway at high speed.

Dissatisfied with the laboratory exercises used in mid-century physics pedagogy, John worked tirelessly on innovative methods that stressed hands-on learning and independent thinking. In 1966, he initiated the Project Lab in which students developed their own open-ended research projects. This is where I had the fortunate opportunity of meeting John during my first semester as an undergraduate at MIT when I enrolled in his Fall 1967 offering of Project Lab.

His belief that anyone could "find something interesting to study about any mundane effect" reflects the independent spirit of John's own early and eclectic science education. He advised his students, "the best way to understand your apparatus is to build it." Charles H. Holbrow, Professor of Physics, emeritus at Colgate University, recalled that John had "the wonderful gift of seeing physics in everyday phenomena and turning these into research projects". Some 2000 fellow MIT undergraduates experienced Project Lab and thankfully this experience has been archived for generations of students to follow. John, along with his co-author Paul Gluck, summarized his many years of mentoring students with "Project Lab" experiments in a book published just after his death in 2014 [8].

John began yet another educational experiment by suggesting and then carrying out "Concentrated Study". He believed that for a significant subset of students, the conventional format of taking a series of courses in different disciplines simultaneously was an inefficient and disruptive way to learn. Rather, he expected a complete and full immersion into one topic at a time, in the style original research is carried out, would be a better way for many students. John convinced the faculty in the other disciplines to join him in offering a sequenced version of the standard MIT undergraduate curriculum.

The model has since been used for many schools for inter-semester immersion courses.

King believed that an understanding of fundamental science concepts should extend beyond physics course curricula. For years he championed the creation of a "Corridor Lab." Never entirely realized at MIT, Corridor Lab would have placed 100 experiments, each demonstrating a scientific principle, along the miles of MIT hallways. Corridor Lab was to be more than a science museum, the experiments were to be calibrated and quantitative. King envisioned similar modules in a wide range of venues to further public understanding of science.

King's most endearing quality as a physicist, witnessed by me, all of his students, and many of his colleagues, was his unbridled enthusiasm for physics and new ideas. One could discuss a concept with him and within minutes drawings and estimates would be made on the back of envelopes and often an interesting and viable experiment was invented on the spot. Ideas did not die when talking to King–the master teacher– they flourished.

References

1. John G. King, *Phys. Rev. Lett.* **5**, 562 (1960)
2. V.W. Hughes, Phys. Rev. **76**, 474(A) (1949)
3. Samuel A. Cohen and John G. King, Nature **222**, 1158–1159 (1969)
4. H.F. Dylla and J.G. King, Phys. Rev. **A7**, 1224 (1973)
5. G. Bressi, G. Carugno, F. Della Valle, G. Galeazzi, G. Ruoso, and G. Sartori, Phys. Rev. A **83**, 052101 (2011); this group of physicists at the Italian National Laboratory (INFN) sent me a preprint of their article that used the same method I used for my neutrality measurement. They pointed out I had made some errors in a detailed calculation that affected the results. I do not think they were amused when they found out I had designed and performed this experiment as an unfunded undergraduate student. All of the equipment I used I managed to find in MIT surplus storerooms except for a modest out-of-pocket expenditure of $3.00 at a local electronic supply store. It is unfortunate for current MIT students that neither storerooms with surplus equipment nor local electronics parts stores are still in existence.
6. H.F. Dylla, J.H. Abrams, C.T. Hovland and J.G. King, Nature 291, 401–404 (1981)
7. An informative summary article on the PSSC project is given by Gilbert C. Finley, *Amer. J. Phys.* **28**, 286 (1960); the PSSC films that have stood the test of time because real physicists were the actors are archived on the following site: PSSC Films-Academic Film Archive, http://www.afana.org/psscfilms.htm

8. John G. King and Paul Gluck, *Physics Project Labs*, Oxford University Press, Oxford, UK (2014)

18

How Long is the Fuse on Fusion?

The popular press occasionally runs stories on the prospects, problems, and opportunities for one of the longest running scientific endeavors—the challenge of producing a practical device that could harness fusion energy as a potentially clean and nearly unlimited source of energy. An example that particularly struck me was a *Washington Post* [1] cover story on fusion energy published in 2012, with a photo showing a physicist peering through the porthole of an experimental magnetic fusion device at Princeton Plasma Physics Lab (PPPL). I began my scientific career at this U.S. Department of Energy (DOE) laboratory in 1975—just two years after a world-wide energy crisis with rationed gasoline and four years before the accident at the Three Mile Island nuclear power plant in Pennsylvania. These events were rallying cries in the U.S. for the development of clean and renewable domestic energy sources.

At that time, fusion energy was little more than a pipe dream with no prospect of contributing to the nation's energy supply in the foreseeable future. Nevertheless, the prospect of the United States leading the development of harnessing the same energy source that powers the Sun helped expand national funding for fusion research during the next decade. Europe, the Soviet Union, and Japan also undertook massive billion-dollar-class demonstration experiments. When the U.S. flagship experiment, the Tokamak Fusion Test Reactor (TFTR) at Princeton was shut down in 1997,

---

Portions of this chapter previously appeared in *Physics Today Online*, 14 March 2012; https://doi.org/10.1063/pt.4.0067.

© Springer Nature Switzerland AG 2020
H. Frederick. Dylla, *Scientific Journeys*,
https://doi.org/10.1007/978-3-030-55800-0_18

the DOE had spent $8 billion (USD) over 45 years on fusion research—more than expenditures on any other energy technology. What do we have to show for this investment? Are we any closer to the day when fusion can contribute to central power generation?

There are no simple answers to these questions. From my career-long look at this endeavor as both a participant and observer, I remain convinced that the goal of demonstrating a sustaining fusion reaction on a laboratory scale remains one of the most challenging scientific feats. We know nuclear fusion works because it is the energy source of the stars and the terrifying engine within hydrogen (thermonuclear) weapons. For use in a practical power station, the fusion reaction has to play out on a scale much smaller than a star and sustained over a much longer time period compared to the flash of a nuclear weapon. The conditions under which a constrained volume of hydrogen fuel generates more energy from fusion than it takes to heat to fusion conditions is called the break-even point. Break-even conditions are reached when a relatively small quantity of hydrogen gas—roughly the amount that would fill a balloon—is heated to the unearthly temperature of 100 million degrees for a time of the order of a second. When I began working at Princeton as a post-doctoral fellow in 1975, my Princeton colleagues working in the field and about a half dozen fusion labs around the world were about a factor of a million away from these break-even conditions. In terms of energy gain, the numbers looked even more daunting—the heated hydrogen gas only produced about a trillionth of the energy required to heat the gas.

Fig. 18.1 Outer edge of the plasma column glowing with infrared light in the PDX fusion experiment at PPPL. The hot dense plasma core emits x-rays not detected by an infrared camera. *Credit* author's photo, PPPL

By 1995, the energy gain had been improved by a factor of 100 billion. And in the mid-1990s, the TFTR at PPPL and the JET Tokamak, a sister experiment in the U.K., performed a series of experiments with the easiest hydrogen isotopes to fuse (a deuterium and tritium mixture) and came within 30–65% of the break-even point! This was a remarkable scientific achievement because it involved a continuous evolution of the design and implementation of solutions needed to better heat, contain, and fuel the hot gas.

I can point to only one other technical endeavor that has shown such a large improvement in key performance parameters. Most people know it as Moore's Law. The continuous improvement of microprocessors is largely due to miniaturization of transistors. Advancements in microelectronics manufacturing technology have enabled the size of a transistor to decrease by about a factor of two every two years. Intel made the first microprocessor chip commercially available in 1971—the Intel 4004 contained 2300 transistors. With careful planning, testing, and implementing technology improvements over the last 5 decades, the Moore's Law road map has held for the microelectronics industry. Intel microprocessors available in 2018 contain more than 10 billion transistors after the company introduced, in early 2017, technology to embed more than 100 million transistors per square millimeter of silicon substrate surface area [2].

But are we any closer to fusion power after the run-up to near break-even demonstrations in the 1990s? A laboratory demonstration of energy break-even is far from the demonstration of a sustained fusion reaction that would generate more energy than it consumed. A practical fusion power plant would need to be engineered to generate excess electrical power for periods of decades with minimal downtime and affordable maintenance. Here is where the promise of fusion energy always appears on the horizon rather than in the short term.

Despite the seemingly large jump from a laboratory scale break-even demonstration to a practical power plant, the next step in fusion energy development is currently under construction in the south of France [3].

ITER, the International Tokamak Engineering Reactor, is being built by an international consortium (European Union, Japan, Russian Federation, United States, Republic of Korea, China, and India) in Caderache, France. The international effort is a sign of the ambition of the project and also a means of sharing the risk and projected cost (greater than $10 billion). This device was originally planned to begin operation by 2016, but initial operations are currently not scheduled until 2026. The primary goal is to demonstrate 500 million watts fusion power for several minutes duration—perhaps a decade after the first experiments. This may sound like a slow, slogging pace when we consider the world's constantly growing demand for energy. The U.S. Department of Energy and numerous international study groups make periodic attempts to construct national and international long-term energy strategies. These plans never appear to be able to influence the entrenched trillion-dollar interests in hydrocarbon-based energy sources. Moreover, with the fiscal realities of energy production and periodic global recessions reining in research budgets, it's a challenge to make a case for funding a long-term solution like fusion research.

Fig. 18.2 The ITER construction site in Caderache, FR (2019). *Photo credit* ITER

This creates a painful reality for governments funding fusion research and the scientists and engineers devoted to this research. Yet I strongly believe that the more than 60-year investment in this "holy grail of energy sources" has made tremendous progress and is positioned to make a major step forward [4]. The U.S. and the other ITER partners should continue to find ways to sustain a critical mass of both domestic and international investments. Despite the remaining challenges, the ultimate goal is too promising to ignore.

The closest analogy for this venture in my mind has a nineteenth century antecedent in America and a medieval antecedent in Europe. For the city of

New York, we have made century-long commitment to construct and maintain very expensive water tunnels that direct water resources from the Catskill Mountains to quench the thirst of the nearly 9 million inhabitants of New York City. The first two tunnels were completed in 1917 and 1935 respectively. The third of these tunnels was started in 1970 and is not scheduled to be operational until the 2020s. It remains the most expensive ever capital improvement project for the city of New York at $6B (USD) and will provide some semblance of reassurance to the city's inhabitants, given the age of the first two tunnels.

In Europe, the cathedral used to be the center of life. In the large cities, cathedral construction often spanned several centuries. I imagine for many who toiled on these monuments, they did not feel they were wasting their time or community resources working on a project that would not come to fruition in their own lifetime. Indeed, it was viewed as a noble calling for the benefit of generations to follow.

That's how I still feel about devoting 15 years of my professional life to fusion research. I hope my experience of watching the field March toward the holy grail of controlled fusion by a factor of 100 billion was not a brief period of raising the cup only to see it fall to the ground within my lifetime. Science has plenty of other problems that will require similar sustained efforts over many decades to achieve a better understanding and possible application. Examples of these other big problems abound in biology, such as the understanding and treatment of cancer, and the visualization of how the brain functions. And for my physics colleagues, there is a recently surfaced problem in cosmology, namely, our lack of understanding of 95% of the contents of the Universe. We give mysterious names to these as yet elusive components of the Universe to hide our current level of ignorance—dark energy and dark matter.

When I was involved with fusion research, I was often asked what motivated me to work on this difficult problem, even though I would probably never see a satisfactory solution in my own lifetime. I didn't have to ponder this question very long. The yearly progress from the 1970s into the 1990s was exciting enough. I was fortunate to cut my scientific teeth on the challenging problem of fusion research. Such scientific endeavors need to be nurtured worldwide, given the scale of investments required for frontier research and the obvious potential for long term benefits to society.

References

1. Brian Vastag, *"Budget cuts threaten the pursuit of nuclear fusion as a clean energy source"*, *Washington Post*, June 25,2012; https://www.washingtonpost.com/national/health-science/budget-cuts-threaten-pursuit-of-nuclear-fusion-as-a-clean-energy-source/2012/06/25/gJQAKlpS2V_story.html?noredirect=on&utm_term=.f04419eed15c
2. Rachel Courtland, *"Intel now packs 100 million transistors in each square millimeter"*, *IEEE Spectrum*, March 30, 2017; https://spectrum.ieee.org/nanoclast/semiconductors/processors/intel-now-packs-100-million-transistors-in-each-square-millimeter
3. The official ITER website is https://www.iter.org
4. For a useful article describing the first 50 years of fusion research see: Dale Meade, *Nuclear Fusion* **50** (2010) 014004; https://doi.org/10.1088/0029-5515/50/1/014004

19

Excitement and Disappointment: The Emotions of Science

Performing frontier scientific research is a very competitive enterprise. Long gone are the days when the world's few scientists had royal or wealthy patrons who funded the scientist's livelihood and research interests. Today's frontier scientific research involves teams of researchers who compete for funds from government agencies, industrial concerns, or private foundations. It is a very competitive business. For frontier research in physics and astronomy, large teams involving hundreds or even thousands of scientists and engineers will work for a decade or more planning and executing a project.

When one reads about the results of a research project either in the standard rote language required for a scientific journal publication or in a more colloquial account in the general press, rarely do such communications capture the emotional highs and lows of the competitive world of modern science.

A major research university is a hotbed of activity for frontier science and is also necessarily the home of the human side of life as a scientist. I was fortunate to spend the first 15 years of my post-graduate career at Princeton University—a research university that was the home of one of the first U.S. academics who could be called a physicist, Joseph Henry [1], and an institution with a long and storied contribution to astronomy, particle physics, and plasma physics.

While reading accounts of two long-standing Princeton research projects in an issue of the science journal, *Nature*, I saw two juxtaposed events that illustrated the emotional highs and lows of scientific discovery. Some of our best science is a product of scientists' persistence and hard work. Sometimes

© Springer Nature Switzerland AG 2020
H. Frederick. Dylla, *Scientific Journeys*,
https://doi.org/10.1007/978-3-030-55800-0_19

scientific discovery is a result of probabilities aligning in a scientist's favor—what some would call plain good luck.

In this particular issue of *Nature* magazine, I learned that a Princeton astronomy team, led by Alicia Soderberg, managed to observe for the first time the initial seconds of an exploding star—a supernova—through the fortunate positioning of an x-ray-observing satellite [2]. Supernovas are not rare events on astronomical time scales; they occur about once a century in any given galaxy. Yet they are usually not detected by earthbound observers until weeks after the stellar explosion, when the first optical light of the explosion appears. The Princeton astronomers' good fortune to catch the x-ray signature of the initial explosion will enable a better understanding of supernovas, which are now understood to be a source of heavy elements in the universe.

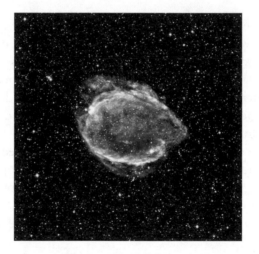

Fig. 19.1 Supernova remnants taken with the Chandra x-ray telescope. *Credit* NASA/CXC/Univ. of Texas

During the same week another group of Princeton scientists was lamenting the cancellation of a major project—the National Compact Stellarator Experiment (NCSX) under construction at the Princeton Plasma Physics Laboratory (PPPL) since 2003 [3]. Since 1952, PPPL has been a flagship U.S. national laboratory for research and development on high-temperature plasmas—gases heated to temperatures of stellar interiors—of more than 10 million degrees. The ultimate goal of the research was to reproduce in practical-sized magnetic confinement devices the conditions of stellar interiors so that hydrogen fusion, the energy source of main sequence stars, could be harnessed. After a half-century of effort by the Princeton team and other groups worldwide, complicated devices called tokamaks in England, the United States, and Japan have approached the conditions necessary to produce net energy output for a few seconds at a time. A large demonstration project was begun in France in 2006 to extend energy production for much longer periods as part of the ITER project [4]. In parallel with ITER design efforts, PPPL scientists and engineers devoted considerable effort to the design and construction of a different type of magnetic bottle. This "compact stellarator" would have portended a simplification of the overall design at the expense of making one particular element—the confining magnets— more complex. Unfortunately, the magnets proved so complicated to design and build that the cost to complete the machine exceeded projections and could not be supported by tightly constrained science budgets. The Princeton team was deeply disappointed that the fruits of their labors are not likely to be completed or tested. They had to regroup and redirect efforts to apply what was learned to existing fusion energy projects. Developing realistic cost projections for unique frontier science experiments is difficult but necessary, especially when it involves the expenditure of public taxpayer funds (See the preceding Chap. 18 for more on the long journey of fusion research).

Fig. 19.2 NCSX fusion device at Princeton Plasma Physics Laboratory (PPPL). *Credit* PPPL

Scientific discovery has its ups and downs. One Princeton team celebrated the good fortune of having its satellite pointed at the right spot at the exact moment a star exploded; another watched in disappointment as funding for its star machine fizzled. One star dies and another is born. With a hooray and a sigh, science marches on.

References

1. Moyer, Albert E., *Joseph Henry: The Rise of an American Scientist,* Washington, Smithsonian Institution Press, 1997
2. A. M. Soderberg, E. Berger, et al *Nature* **453**, 469–474 (2008)
3. Statement by Raymond L. Orbach, U.S. Department of Energy on the cancellation of the National Compact Stellarator Experiment (NCSX) at Princeton Plasma Physics Laboratory, May 22, 2008; https://ncsx.pppl.gov//DOE_NCSX_052208.pdf
4. ITER Project. see https://www.iter.org

20

Considering Future Energy Options: Extrapolations from a Real Experiment

About once a year I look forward to spending some time with a good friend, a fellow physicist with whom I shared nearly 45 years of collaboration and conversations, largely around the subject of energy. My friend is Fritz Wagner, a former director at the Max Planck Institute of Plasma Physics (IPP) in Greifswald, Germany, and a recognized international expert in plasma physics, experimental fusion devices, and analysis of the world's current energy situation.

Since Wagner retired as IPP director in 2008, he has spent much of his post-retirement years studying the global energy situation and potential solutions. Given that the economic health of a developed nation is proportional to the availability and use of energy resources, this subject is clearly connected to our well-being. Many factors influence this issue including the often competing demands of economically efficient energy production, control of CO_2 (carbon dioxide) release, and the global interconnectivity of energy production and use. A singular goal would be to minimize the all-too-frequent conflicts over the acquisition and distribution of hydrocarbon fuels.

Wagner published an important study on an optimal mix of renewable and conventional energy sources based on a unique and ongoing experiment in his home country of Germany. Germany is not only the economic powerhouse of the European Union; it is also the world's renewable energy kingpin. Germany's sophisticated energy infrastructure and its interaction with the rest of the EU power grid draws attention to the need for balancing energy production with demand, and for achieving stability in the power transmission interconnections across political boundaries.

© Springer Nature Switzerland AG 2020
H. Frederick. Dylla, *Scientific Journeys*,
https://doi.org/10.1007/978-3-030-55800-0_20

Why is Germany so interesting as a test case for these important problems? With an installed renewable energy (RE) production capability at the end of 2018 of 45 GW photovoltaic (PV) and 59 GW wind, the total renewable energy production now exceeds the typical peak electrical load (80 GW) for the entire country. Since there is very little energy storage capability in Germany (nor anywhere else for that matter), the peak energy production from RE generation sources must be shipped to other countries through the European grid, and the production level of existing thermal energy plants must be turned down. Transporting the power to other consumers on the EU grid often leads to a situation where Germany is paying other countries to absorb the power—leading to the strange situation of negative electricity prices. The alternative of turning down a thermal plant is restricted due to technical and economic considerations. Such plants basically operate between two limits, and the start-up and turn-off costs are high because large thermal plants are not designed for intermittent use. The problem of localized energy production exceeding demand could also be dealt with if technology existed for large-scale energy storage. Presently, in Germany, the existing storage capacity based on pumped hydroelectric facilities is only 40 GWh, which is a factor of 500 less than is needed in an ideal situation for seasonal load balancing with the present demand curves. In addition, the prospects for any new high-capacity energy storage capability based on batteries or chemical conversion technologies are far from near-term use, with respect to both scalability and cost.

Fig. 20.1 Wind turbine farm in Bernburg, Germany. *Credit* Franzfoto, Wikipedia Commons, CC-BY-SA-3.0

Wagner published his analysis of the present energy situation in Germany in several scientific journal articles [1–3]. He outlines an option for both Germany and the rest of the European Union that entails optimizing the mix of renewable energies versus conventional sources, and taking full advantage of strengthened interconnections within the EU-wide power grid. The analysis attempted to minimize the necessary backup needs, with consequent reductions in the required storage capacities and CO_2 production. Wagner's analysis shows that there is a limit for electricity contribution of RE of about 40% of the residual annual base load. Beyond this limit, grid power and surplus production increase, and in addition, there is an optimal mix of PV (photovoltaics) to wind-produced energy because of the different time characteristics of their production principles. By balancing the geographic variations across the very different environments of northern and southern Europe, overall variations are minimized in a case optimized for the entire continent.

Given that there are no near-term prospects for scalable storage capacity, other options are possible—such as using price variability to influence the night/day demand variations. Somewhat surprising is the fact that, with the optimal installed RE capability, daytime becomes the time of greatest production capability and thus offers the lower cost of delivery to consumers. Larger scale use of electric cars may offer some storage capability, and given the need for a roughly equal capacity for large-scale thermal plants, options for heat cogeneration and electrolytic generation of hydrogen or methane may also be viable options.

This analysis could be extended to North America. The continent has large areas suitable for PV plants (the American West) and significant capability for wind generation. Moreover, there is a much more economically favorable conventional generation capability based on the recent exploitation of natural gas captured by hydraulic fracturing of shale gas reserves. However, it is difficult to imagine that the present U.S. political environment would yield anything like the powerful tax benefits offered by the German government to propel the nation forward to become a world leader in renewable energy. The basis for nuclear-based energy generation is zeroed out in Wagner's analysis, given the post-Fukushima political liabilities and the subsequent actions by the German government to phase out power generated by existing German fission plants by 2022. The other nuclear option (fusion) is part of neither the near-term nor the mid-term equation.

Both Dr. Wagner and I began our scientific careers in fusion research, and Wagner spent his entire professional career working in this frontier field of research. We both agree that for inclusion in any analysis of energy generation, this technology is not part of the world's energy sources for the foreseeable future. However, its prospects to minimize the waste and fuel cycle concerns of fission and to avoid the deficiencies of intermittent electricity sources—which are becoming increasingly obvious from Germany's "real experiment"—are so great that worldwide research efforts on this energy frontier should definitely be continued, if not intensified.

References

1. Friedrich Wagner, "*Electricity by intermittent sources: An analysis based on the German situation 2012*" Eur. Phys. J. Plus **129**, 20 (2014); https://doi.org/10.1140/epjp/i2014-14020-8
2. Friedrich Wagner, "*Considerations for an EU-wide use of renewable energies for electricity generation*" Eur. Phys. J. Plus **129**, 219 (2014); https://doi.org/10.1140/epjp/i2014-14219-7
3. Friedrich Wagner, "*Surplus from and storage of electricity generated by intermittent sources*" Eur. Phys. J. Plus **131**, 445 (2016); https://doi.org/10.1140/epjp/i2016-16445-3

21

My Encounters with the Queen of Carbon

I was fortunate to spend most of scientific life knowing Millie Dresselhaus. I began my undergraduate education in physics at MIT in the year (1968) that Millie was promoted to a full professor—the first woman given that honor. I recall hoping that distinction would become less of a rarity than I observed in my undergraduate class of nearly 100 fellow physics students that was enriched with only 10 very bright women—still a small fraction, but nearly 10 times higher than the statistics at other US colleges and universities at the time.

I first became aware of Millie's growing reputation as an expert on all forms of the element carbon when I began my PhD research in 1971. I spent the next four years developing a new kind of electron microscope that could image delicate biological samples without causing damage to the structure and function of the sample. It turned out I needed a special form of carbon in my samples to test the technique [1]. Having the world's expert and her students in the next building at MIT solved my problem.

After graduating from MIT, I spent the next 15 years of my career at Princeton University's Plasma Physics Laboratory as a materials scientist hired to work on the difficult "first wall" problem for magnetic fusion devices. Three large "tokamak" machines were under design and construction at Princeton when I arrived as a fresh post-doctoral fellow in 1975. Over the

Note Previous abridged versions of this article appeared in *Physics Today Online, May 23, 2017* (published by the American institute of Physics), and in the Sept. 2017 issue of *Technology Review* (published by MIT).

© Springer Nature Switzerland AG 2020
H. Frederick. Dylla, *Scientific Journeys*,
https://doi.org/10.1007/978-3-030-55800-0_21

next 15 years, I found myself returning to Millie's research on carbon as I helped develop techniques for removing unwanted carbon impurities from fusion research vessel walls [2] or designing specific forms of carbon that could withstand exposure to high heat fluxes [3].

In 1990, the next stop in my career was joining the scientific staff at Jefferson Lab, a U.S. Department of Energy accelerator facility in Newport News, Virginia. I found myself in the thick of helping to design and build two large accelerators that required me to call on Millie's continued portfolio of expertise on various forms of carbon. One form of carbon capable of absorbing radiofrequency energy at superconducting temperatures became a small magic bullet that enabled the operation of the Jefferson Lab accelerators and all superconducting accelerators that have since been built worldwide. In the mid-term of my stint at Jefferson Lab, I was exposed to her steady and knowledgeable hand as an administrator when she was appointed Director of the Office of Science for the Department of Energy. She had to deal with the complicated budgets and politics of the Department of Energy's national laboratories and myriad university research programs, but she still found time to devote to her MIT students. For the first time, I was involved with research that directly overlapped her current interests—a unique form of carbon called nanotubes. These small tubes with widths of atomic dimensions would turn out to have wide application to electronics and high strength structures. My colleagues at Jefferson Lab were making more of a purified form of these nanotubes [4] than anyone else at this time by using the Jefferson Lab's powerful free electron laser facility [5].

The next stop in my career brought me even closer to Millie's wide sphere of influence. In 2007, Millie was in the middle of a 5-year term as Chair of the Governing Board of the American Institute of Physics (AIP)—a federation of ten scientific societies representing more than 135,000 scientists and engineers. In that year, I was selected by Millie and her Board as AIP's next Executive Director. This is when I really got to know Millie up close and find out what a remarkable person she was. Her love of science, her dedication to teaching, and working with students of any age (including me in my late 50's), and her ability to wrangle through knotty administrative problems were on full display. During quiet moments over a shared cup of tea, I was exposed to her equal love for music and the numerous opportunities she found in her busy life to share music making and appreciation with her family. I soon missed the adventure of frequent interactions with Millie when she completed her term as AIP's Board Chair in 2009. But we stayed in touch through encounters at scientific meetings and emails on shared interests that continued to grow since my first encounter with the "Queen of

Carbon" in 1968. My colleagues and I at AIP were enthusiastic supporters of Millie's nomination for the National Medal of Freedom that she was so rightly awarded in 2014.

Fig. 21.1 Mildred Dresselhaus. *Credit* Michael D. Duncan, courtesy AIP Emilio Segrè Visual Archives

I miss that spunky lady with her signature red sweater and ready smile. The scientific community lost Millie in February 2017 after more than 50 years on the faculty at MIT and inspiring generations of students and collaborators. She was still working with her students, post-doctoral fellows, and scientific colleagues right up until the day she left this world [6].

References

1. H.F. Dylla, J.H. Abrams, C.T. Hovland and J.G. King, *Nature* **291**, 401–404 (1981)
2. H.F. Dylla, *Journal of Vacuum Science & Technology* A **6**, 1276 (1988); https://doi.org/10.1116/1.575689
3. J.L.Cecchi et al, *J. Nucl. Mater.* **128** (1984), 1–9; https://doi.org/10.1016/0022-3115(84)90322-2
4. P. C. Eklund, B. K. Pradhan, U. J. Kim, Q. Xiong, J. E. Fischer, A. D. Friedman, B. C. Holloway, K. Jordan and M. W. Smith, *Nano Letters* **2**, 561, 2002; https://doi.org/10.1021/nl025515y
5. H. Frederick Dylla and Steven T. Corneliussen, *"Free-Electron Lasers Come of Age"*, *Photonics Spectra*, p. 65, August 2005; G.R. Neil, C. Behre, S.V. Benson,

M. Bevins, G. Biallas, et al, *Nucl. Instr. & Methods* **A557** 9 (2006); https://doi.org/10.1016/j.nima.2005.10.047

6. Millie Dresselhaus's obituary article in the *MIT News*; http://news.mit.edu/2017/symposium-commemorates-pioneering-professor-mentor-mildred-dresselhaus-1129

22

A Prize Hidden Under a Piece of Tape

The Nobel Prize award announcements in the first week of October every year are always eagerly anticipated by the scientific community. As a physicist, I always paid particular attention to the physics prize. Given that this prize is considered the most prestigious prize in science, it is often bestowed on a scientist or a trio of scientists whose discoveries or inventions have stood up to a career-long examination by the physics community. When the Nobel Prize is awarded to a young scientist, it is particularly noteworthy. That was the case for 2010 when the Nobel Prize in Physics [1] was awarded went to Andre Geim and Konstantin Novoselov, both from the University of Manchester. Their award, for the discovery and investigation of graphene, a novel two-dimensional form of carbon, is noteworthy not only because of the investigators' young age, but also since it is rare for a Nobel Prize to be awarded so soon after the prize-winning work—in this case the initial discoveries were in 2004.

The 2010 prize recognizes a very special form of carbon—a crystalline layer only one carbon atom thick–very difficult to isolate even though it can be made from the common form of carbon we see every day in pencil lead or flame soot. The prize winners learned to isolate graphene flakes and used the samples to demonstrate its unique properties. Two of those properties promise an array of practical applications. Graphene, as a near-perfect conductor of electricity at room temperature, portends a new family of microelectronics with faster switching speeds and less power consumption. Also, as one of the strongest materials known, graphene could be embedded in a new class of composite materials to strengthen them.

© Springer Nature Switzerland AG 2020

H. Frederick. Dylla, *Scientific Journeys*,

https://doi.org/10.1007/978-3-030-55800-0_22

The work leading up to this prize is an example of the scientific discovery process in full bloom. In 1996 Richard Smalley, Robert Curl, and Harold Kroto won the Nobel Prize in Chemistry [2] for their discovery of a related form of carbon, fullerenes. Although widely available in nature, fullerenes were never isolated until the 1980s when those men published papers on the soccer-ball-shaped conformations of carbon more affectionately called "buckyballs," by reference to the geodesic domes pioneered by architect Buckminster Fuller.

In the past two decades yet another form of carbon has become famous. Scientists learned to produce tiny tubes of carbon only a nanometer across. Those carbon nanotubes burst on the scene, launching an entirely new field of study owing to their interesting physical and chemical properties. Possible near-term applications are expected in electronics, pharmaceuticals, and the development of new materials [3].

Graphene, nanotubes, and buckyballs are all related. Unwrapping a buckyball or splitting open a single-walled nanotube generates a small sheet of graphene. But there was no reproducible way of making graphene until Geim and Novoselov combined sophisticated micro-handling techniques with the dexterous use of Scotch tape.

In legislative bodies, grant funding agencies, and corporate R&D laboratories, the relative merits of funding basic versus applied research are often debated. I have always been uncomfortable with these labels. The isolation of graphene is a good example of the narrow gulf between basic and applied research. Here we can plainly see that fundamental physics and the development of useful devices are not separate interests but intimately related. It is precisely graphene's strange quantum properties—including the fact that an electron's speed through graphene is the same no matter how much energy the electron has—that might lead to smaller, cheaper, and faster electronics, helping to extend Moore's law (according to which computing power doubles about every 18 months) for many years to come.

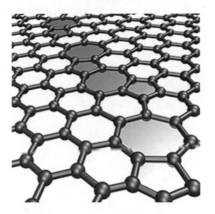

Fig. 22.1 Structure of a graphene sheet. Image courtesy of Oleg V. Yazyev and Steven G. Louie, UC Berkeley and LBNL

The 2010 Nobel Prize winners did not launch their scientific investigation of flaky carbon in order to build a better switch or light bulb. Instead they were curious about the ability to isolate this ubiquitous but long-hidden form of carbon. Now nearly two decades later, the fruits of their curiosity are more abundant than they could have dreamt—not just in the recognition of the Nobel committee, but in all the new switches, materials, and light sources in our future that will have graphene at their core.

References

1. The Nobel Prize in Physics Citation for 2010; https://www.nobelprize.org/prizes/physics/2010/summary/
2. The Nobel in Chemistry Citation for 1996; https://www.nobelprize.org/prizes/chemistry/1996/summary/
3. Michael F.L. DeVolder et al,"*Carbon nanotubes: present and future applications*", *Science* **339**, 535–539 (2013)

23

The Big Machine

For physicists and many who have casual interests in frontier science, one of the most remarkable machines ever built for fundamental research is the Large Hadron Collider (LHC), located in a tunnel deep beneath the Jura Mountains outside of Geneva, Switzerland.

The machine is the largest particle accelerator ever built by the international high-energy physics community. Its completion and successful commissioning in 2008 constituted the culmination of a decade-long, $5 billion (USD) project, funded by more than 30 countries, at the European Organization for Nuclear Research (CERN), headquartered in Geneva.

Fig. 23.1 The Large Hadron Collider at CERN. (*Credit* CERN)

For more than a century, physicists have used directed beams of subatomic particles to explore the fundamental constituents of the physical world. Indeed, the notion of particles smaller than atoms was confirmed by J. J. Thomson, who discovered the electron in his Cambridge University laboratory in 1897. Thomson produced a primitive beam of electrons in a glass tube no bigger than the first television receiver tubes. Several years later Ernest

© Springer Nature Switzerland AG 2020
H. Frederick. Dylla, *Scientific Journeys*,
https://doi.org/10.1007/978-3-030-55800-0_23

Rutherford discovered the atomic nucleus in the same laboratory. Rutherford measured the deflection through a sheet of gold foil using a beam of alpha particles (each a tiny parcel consisting of two protons and two neutrons) emitted by a radium sample (see Chap. 3, for more on Rutherford). From those primitive experiments, physicists have designed, built, and employed a series of "atom smashers," or particle accelerators, over the last century to explore the constituents of matter—a whole family of subatomic particles discovered since Thomson's time. The machines have also given us a glimpse of the early Universe evolving from the Big Bang about 13.7 billion years ago.

The LHC is no doubt one of the most complex machines ever built [1]. The accelerator requires 1600 precision superconducting magnets, cooled to near absolute zero (1.8 K) to conserve power, along the 27 km beam path. In the heart of this machine are house-sized detector assemblies that straddle the beamlines at two strategic locations, where two counterrotating beams are guided to collision points, enabling effective collision energies of up to 7 teravolts. The infrastructure required for the collider to collect, distribute, and analyze the experiments, which began in 2009, is as impressive as the machine hardware. The infrastructure for the two large detector assemblies, called ATLAS and CMS [3], was under development for the same length of time as the accelerator. More than 5000 scientists, from several hundred research institutions and dozens of countries, have put together the largest grid computing network this side of the Googleplex to analyze the LHC data over the nominal quarter century life of the machine in its present incarnation. The architecture behind the grid computing network is just one of the technical spinoffs from the development of the LHC that will benefit many other fields of science and commerce. The LHC generates about 15 petabytes (one million gigabytes) of data per operational year, the equivalent of a stack of CDs 20 km high. These data are shared over this network with thousands of scientists worldwide to facilitate the difficult task of identifying and analyzing the key signatures of new subatomic particles and physics among the minute fraction of the collected data that is important—it is literally a problem of finding a needle in a haystack.

Why should anyone beyond the high-energy physics community care about the existence of this mega science project? The many answers to this question span the economics of frontier science, the politics of international collaborations, the engineering of unique hardware, and the nature of information networks that can handle the envisioned data loads [2]. When colleagues from the other sciences or audience members cringe after I have given public talks over the costs of high energy physics experiments—overwhelmingly funded by the public through taxes—I simply point to one

"spin-off" technology from CERN that was given to the rest of the world—
the World Wide Web. CERN physicist, Tim Berners-Lee is well recognized
and heralded as the inventor of the Web [6]. This now universal means of
transmitting and displaying large quantities of data was invented and developed in the early 1990s by Berners-Lee and his colleagues at CERN in the
design phase of the LHC to solve the problem of efficiently transmitting,
displaying, and interacting with CERN's data flows throughout its growing
worldwide network. Several years later (1994–95), the first commercial web
browsers were introduced. Combined with the surge in deployment of low–
loss optical fibers that interconnected the entire planet by the turn of the
century, the web became the communications hub for all world commerce.
It is difficult to put a monetary value on that one gift to the world by CERN
and the high energy physics community. Clearly, if a very small royalty were
attached to each data packet transmitted on the web, no additional public
funding would be required for basic research in any science for many years to
come.

There are many spin-off technologies that have benefited from the entire
worldwide effort of designing and building particle accelerators and the
detector infrastructure needed to make these devices useful for science.
Because these projects are so large and complex—indeed they were the first
projects to be given the moniker "Big Science"—their design and construction necessarily involve work in the forefront of many technologies. Common
examples beyond computing include the first large scale superconducting
magnets that are now available in every MRI imaging device, sensitive and
cost-effective detectors for medical imaging (used in CT and PET scanners),
and small accelerators available in regional medical centers for cancer therapy.

Even though I spent a good portion of my career building large accelerators
and magnetic fusion devices, I never felt comfortable justifying to the public
that the value of a large expenditure of public funds on a science project was
the potential for spin-off technologies. Why not just spend the money on
the spin-off technology and not bother with the expensive project that may
simply satisfy the curiosity of a few thousand physicists?

Because such questions concerning the value of basic research still arise in
discussions of the public funding of science, the scientific community still
has not done enough to show that basic research pays off in many ways—the
relevant timescale is just not the next quarter or year.

[I offer more commentary on this subject in Part 3.] But let's return to the
value of the LHC.

To me, the most important result of the construction of the LHC and
indeed the existence of CERN, the remarkable multi-international laboratory

that houses this project, is its social value. CERN was founded in 1953 with the clearly expressed goal of relaunching European science after the devastation of World War II. The original laboratory was founded with 12 charter member countries from Europe. Both its management model and funding model were unique and have passed the test of a half century of success. CERN is managed by a multi-national council comprised of CERN members with each member making an agreed annual budget contribution. From its founding days, the council has grown to 23 member and 8 associate member countries (as of 2019). Membership permits that country's participation in CERN's activities. The model works—over its first 50 years, a succession of accelerators has been built and operated primarily for the benefit of nuclear and particle physics, but along this journey significant contributions have been made to astrophysics, atomic physics, and medical physics. Many of the early accelerators are still operating as precursor devices that first generate the protons and initially accelerate them before injection into the large LHC ring. The myriad of design and construction efforts in order to build a state-of-the-art particle accelerator have led to consistent nurturing of engineering expertise in materials science, metallurgy, vacuum technology, magnet design, electronics, and computer technology that is envied worldwide. The CERN management model has ensured stable funding and cross-continent development of industrial partners. Many national and international "big science" projects have mimicked the CERN model or wish they had adopted it from the beginning.

So, the LHC cost about $5 billion (USD) to design and build and an additional expenditure of nearly $2 billion (USD) was spent for the detector assemblies and computing infrastructure. The key scientific mission of the LHC was achieved in 2012 when CERN announced the discovery of the Higgs Boson—the missing subatomic particle that was first predicted by Peter Higgs from the University of Edinburgh and independently by Francois Englert and Robert Braut at the University of Brussels in the mid-1960s to explain the fundamental nature of mass [5]. Higgs and Englert met in person for the first time at CERN when the Higgs discovery was announced. A year later they were awarded the Nobel Prize in Physics. It is notable that award citation gave credit to the 5000+ members of the scientific teams from the CERN Atlas and CMS detector groups for the discovery of the particle [4]. The LHC will continue operating at least through the mid 2030s with a number of upgrades. The LHC may or may not find signs of things we've never seen before, such as supersymmetric particles or extra-spatial dimensions beyond the three dimensions of everyday life. That is the nature of basic scientific research—very few frontier discoveries are predicted.

By its very existence, CERN has already justified its and LHC's existence many times over. I can think of no better example of a successful multinational project that encourages so many cultures and nationalities to work together. The over 10,000 scientists and engineers from over 100 countries who work and interact with CERN is proof of that assertion. This includes representatives from nations who in any other context would not even talk to each other let alone work together. CERN is the best example of money well spent to get the world to work together in harmony on a shared goal. Science is good for that above all else.

References

1. An excellent reference on the design and construction of the LHC by CERN's Director General who oversaw much of the effort is given by C. Llewellyn-Smith, *The Large Hadron Collider, Scientific American,* July 2000.
2. A comprehensive economic and social benefit analysis of large international science projects with a specific focus on LHC is given by Massimo Florio and Chiara Pancotti, *The economics of physics: A social cost-benefit analysis of large research infrastructures, Oxford Research Encyclopedia of Physics,* Oxford UK, 2020; see https://doi.org/10.1093/acrefore/9780190871994.013.23
3. Jon Butterworth, a lead physicist with the ATLAS collaboration at LHC, describes the nearly 20-year venture of design, building and using the ATLAS detector for LHC's most prominent discovery to date—the Higgs Boson: Jon Butterworth, *Most wanted particle,* The Experiment, LLC New York, 2015.
4. The 2013 Nobel Prize in Physics was awarded to Peter Higgs and Francois Englert, NobelPrize.org. https://www.nobelprize.org/prizes/physics/2013/summary/
5. There are several excellent books on the discovery of the Higgs written by physicists with a gift for exposition of science to the general public. See for example: Lisa Randall, *The power of empty space,* Bodley Head (2013), and Sean Carroll, *The particle at the end of the universe,* Dutton (2012)
6. Tim Berners-Lee, *Weaving the Web,* Harpers, San Francisco (1999). Berners Lee has been the Director of the World Web Consortium since 1994; his biography on the WC3 site gives many useful references on the history and subsequent development of the web: https://www.w3.org/People/Berners-Lee/

24

The Collider that Couldn't

"Super" is an apt description for the Superconducting Super Collider (SSC) project, the ill-fated particle accelerator that was well under construction near Dallas, Texas, before it was cancelled by the US Congress in 1993. The project was super in size and energy—84 km in circumference, 20 trillion electron volts (TeV) per beam—and it would have been the largest single application of superconducting magnets. If completed, the SSC would have been the most powerful device of its kind, with more than double the energy of CERN's Large Hadron Collider.

However, the SSC got caught in a super mess: small science pitted against big science, pork-barrel politics, conflicting management styles, and changing views of the value of publicly financed investments in science. Ultimately, the downfall of the SSC project was in large part due to rising construction costs, which made the project's budget estimate a moving target: In 1986 it was $3 (USD) billion; by 1990 it was $8–$12 billion. The escalating cost became the story in popular and scientific publications during a time when the U.S. Congress was looking for spending cuts to offset rising federal deficits.

The prospect of siting a multibillion-dollar capital project with a $330 million annual operating budget and more than 1000 permanent staff was a golden prize when the site selection process began in 1988. The project's sponsoring agency, the Department of Energy, chose Waxahachie, Texas—just outside the Dallas suburbs—as the winning site from among 43 bids.

This section was adapted from a book review first published by the author in *Physics Today*, March 2016, **69** (3) (AIP, Melville, NY, 2016).

© Springer Nature Switzerland AG 2020
H. Frederick. Dylla, *Scientific Journeys*,
https://doi.org/10.1007/978-3-030-55800-0_24

Tunnel Visions: The Rise and Fall of the Superconducting Super Collider [1], a nearly three-decade writing project, describes the birth and death of the SSC. The authors—Michael Riordan and Lillian Hoddeson, both science historians, and Adrienne Kolb, a retired Fermilab archivist—illuminate the serious problems that led to the 1993 congressional vote to terminate the SSC. Those included a faulty transition in expertise from the project's original Central Design Group to the SSC Laboratory that resulted in the inability to attract and retain seasoned accelerator design staff, not to mention several failures in project management.

The painful history of management pitfalls is laid bare through the authors' extensive interviews and perusals of official government records. For example, the Department of Energy (DOE) allowed just three months for contracting teams to pull together a management proposal that included industrial partners. Only the Universities Research Association, which managed Fermilab, was able to respond fast enough. Over time, an increasingly dysfunctional relationship between project managers and DOE administrators responsible for oversight resulted in a cumbersome structure of four poorly communicating levels of management—two from the association and two from DOE.

Riordan, Hoddeson, and Kolb detail the project evolution from conceptual design to establishment of the SSC Laboratory to a series of increasingly layered management structures whose intricacies hastened the project's termination. Changes from the original design increased the size and complexity of the magnets, which led to a concurrent doubling of the front-end injector energy. Those changes would have accelerated commissioning and provided more confidence in attaining the full design energy—but they also added $2 billion to the cost. The authors suggest that it would have been better to site the SSC at Fermilab and to copy the CERN model, in which a series of new and larger accelerators could concatenate using existing infrastructure and existing staff.

The accelerator's escalating cost was a ripe target for certain members of Congress—for example, the New York delegation, which was already smarting both from losing the SSC site competition and from DOE's 1983 decision to cancel the Isabelle accelerator at Brookhaven National Laboratory. As the SSC's costs rose, other segments of the scientific community also targeted it. If the SSC were to be completed and operated for several decades, opponents claimed, there would be less money for federal funding of the rest of science. The materials-science community was particularly vocal.

In the end, the House voted 282–143 to cancel the project. The U.S. high-energy physics community never fully recovered.

Fig. 24.1 Tunnel under construction in 1993 for the SSC accelerator. *Credit* Fermilab archives

'I observed the political drama while managing the construction of another superconducting accelerator at the Continuous Electron Beam Accelerator Facility, part of DOE's Thomas Jefferson National Accelerator Facility (Jefferson Lab) in Newport News, Virginia. That was a smaller project, but it featured aspects of the SSC project management model and executed them well. It had a strong, hierarchal management scheme, had ties to the national and international research communities, conducted outreach to fields other than nuclear physics that could use the laboratory's infrastructure, and carefully cultivated political support. Because of my personal ties to the field and to the national lab community, I enjoyed the play-by-play account in *Tunnel Visions*. Others without such connections may dispassionately read this retelling of the birth and death of a megaproject. The SSC has lessons for all who advocate the public funding of science.

Reference

1. Michael Riordan, Lillian Hoddeson, and Adrienne W. Kolb, *Tunnel Visions: The Rise and Fall of the Supercollider* (U. Chicago Press, 2015).

25

One Man Can Make a Difference

Building a complicated and expensive machine for studying the frontiers of physics is far from a conventional construction project. All such machines, including particle accelerators, large telescopes, and robotic space satellites, usually require more than a decade of planning and design studies. Strong scientific leadership is needed to organize the scientific user community who operate and benefit from these instruments and to craft the message to advocate to fundraisers. Since these instruments are dedicated to scientific discovery and basic research, the vast majority of the construction and operational funds are provided by public funds. An absolutely essential component of building a big scientific instrument is convincing funding agencies and taxpayers that the project is worthy of a large expenditure.

This is an abbreviated account of how one large machine was built in Virginia, largely due to the efforts of one unique individual, Hermann Grunder [1]. I had the pleasure of working for this man during the 1990s—a decade during which he oversaw the successful construction and operation of a large electron accelerator with the awkward name of CEBAF, an acronym for the Continuous Electron Beam Accelerator Facility. Of equal importance to the appearance of this new science tool was the coalition of local, regional, national, and international supporters he put together for the project. The project became invaluable to a cadre of international nuclear physicists who have used the machine for more than two decades for studies advancing our still incomplete understanding of the subatomic world. What transcended the original scientific utility of this machine was the value Grunder created by building a new national laboratory surrounding the electron accelerator. This

© Springer Nature Switzerland AG 2020
H. Frederick. Dylla, *Scientific Journeys*,
https://doi.org/10.1007/978-3-030-55800-0_25

laboratory and its staff provided an educational resource for local school children, colleges, and universities, and industries requiring access to technology and unique infrastructure (see Chap. 26).

The project started out as most so-called "big science" projects started in the developed world in the latter half of the twentieth century. First, an interest group of scientists proposed a next-generation machine to advance the state of the art. In this case, a group of nuclear physicists in the U.S. proposed a new electron accelerator to replace outdated machines. The project was designated the National Electron Accelerator Laboratory (NEAL). It was to provide a high current, multi-billion-volt electron beam to probe the structure of the atomic nucleus.

Physicists have been building and using particle accelerators since the time of Ernest Rutherford, who discovered the nucleus in 1911 by using high-energy (alpha) particles that are naturally emitted by the element radium (see Chap. 3). For more than a century, physicists have designed increasingly larger and more powerful accelerators to study the basic constituents of matter. The nuclear physics community was particularly interested in using high energy and tightly focused electron beams to probe the inner structure of the individual particles that make up the nuclei of all atoms—protons and neutrons. The higher the energy of the electron beam, the smaller the volume of a proton that could examined. By increasing the electron beam current, the minute signals from these microscopic investigations could be magnified. In the 1980s the nuclear physics community started asking for proposals for the specified electron accelerator and an array of large sensitive detectors as the "eyes" of the experimental facility.

As is the custom for large scientific projects, proposals were solicited by the relevant funding agency (in this case the U.S. Department of Energy) for the design and preliminary cost estimates for the new accelerator. Proposals for building state-of-the art accelerators at the time were crossing the $200 (USD) million price tag, which meant that multiple laboratories were eager to host such a new project. Proposals were submitted from the traditional U.S. national laboratories and large research universities. A modest collaboration from the University of Virginia Physics Department—a newcomer to make such a proposal and from a newcomer region of the country—submitted a proposal that unexpectedly survived three rounds of national peer review for being the best project. Before the project was given a preliminary go-ahead, the Department of Energy required the Virginia team to strengthen their proposal team in two essential ways: expand the consortium beyond one research university to ensure a larger talent pool of machine builders and nuclear physicists, and hire a credentialed project director who had experience in successfully building machines of this scale. The first demand was relatively easy to satisfy. There were few "big science" projects in the Southeastern U.S. In short order the Virginia team built a consortium of research universities than included all of the research universities in Virginia and many in the Southeastern U.S. The second demand of securing a seasoned laboratory director required more effort.

This search included wooing of a kind to which scientists are rarely exposed. The Governor of Virginia sent his plane to Berkeley, CA, to bring Dr. Hermann Grunder on a tour of the state and a largely wooded 200-acre site in Newport News, Virginia, that would become the site of the new accelerator laboratory. Grunder, at the time of his courting, was the Associate Director in charge of accelerators and fusion studies at the Lawrence Berkeley National Laboratory—the nation's first accelerator laboratory housing a progression of accelerators inspired by Ernest Lawrence since his first Nobel-prize-winning cyclotron in the 1930s.

Fig. 25.1 Hermann A. Grunder. *Credit* Jefferson Lab

Grunder was born and educated as a physicist and engineer in Switzerland and became an American citizen during his early days working at Berkeley. By the 1980s, he was a sought-after expert in the management of big science projects because of his combination of scientific, engineering, managerial, and political skills. The Virginia consortium could not have found a better person, not only to build the new accelerator, but also to build a new national laboratory that would serve as the host institution for what would become a significant educational resource for the region and one of three major centers for nuclear physics in the U.S. over the next few decades.

When Grunder first saw the laboratory site in 1985 there was only a small office building and one gymnasium-sized concrete structure that had formerly housed a small proton accelerator, used by NASA in the 1960s. Even before Grunder accepted the job as lab director he sampled a range of professional opinions about the proposed project, including physicists from the winning design and those from competing design teams. He also carefully evaluated the prevailing opinions of the federal officials at numerous government offices and within the U.S. Congress who would have to approve the project and its budget before it received official approval for construction. Grunder seriously considered not accepting the job as lab director because of two serious concerns. First, there were reasonably high political uncertainties that the project might not be approved. Second, after his evaluation of alternative technologies he became convinced that the design approved for the new machine was too conservative for a major new scientific project that should have a lifetime of several decades. Politicians and bureaucrats who must approve budgets are typically wedded to conservative projections for complicated projects; however, Grunder became convinced that this project could be a platform to demonstrate a new technology that would have wide application beyond this one project.

In 1985, Grunder accepted the job to build this new scientific facility. The approved preliminary design called for an electron accelerator based on well-established accelerator structures made of copper that feed radio frequency power for accelerating the electron beam (many of us have a miniature version of this structure in our microwave ovens). The design was approved, the budget was established ($236 million), and a timeline was set in place for getting the project of the ground, with the first construction funding approved for 1987. Everything seemed ready for take-off, except Grunder became convinced that he did not want to launch a new experimental facility on the basis of an established technology that would be difficult to upgrade to higher energies as the science progressed. During this tenuous pre-project approval phase, Grunder made a worldwide survey of a next-generation accelerator technology based on superconducting radio frequency cavities—SRF for short. There were pockets of expertise using these cavities at Stanford and Cornell Universities in the U.S., and in several laboratories in Germany. At that time only a few of these cavities had been manufactured and tested, but Grunder had to convince a long list of "approvers" that more than 300 could be made reliably for the CEBAF machine, in addition to producing a completely new accelerator design that would use these structures instead of the approved copper structures that operated at room temperature.

A key point in the decision-making process occurred in a project review with external scientific reviewers invited by the Department of Energy. Sensing the reviewers' concerns for the risk of moving to an unproven technology, combined with the federal government representatives' nearly toxic fears for any changes that would affect the project cost estimate, Grunder asked a representative of one of the companies who had offered to make the required number of superconducting cavities whether they would accept a fixed price contract for the job. Without excessive hesitation, the vendor said he would. This was one of many examples of Grunder's uncanny ability to ask the right question at the right time to sway an audience when there were technical or political uncertainties.

Fig. 25.2 Superconducting RF (SRF) accelerator cavities under test in a clean room. *Credit* Jefferson Lab

The CEBAF project was approved for construction in 1988 with the new superconducting technology—with the government's proviso that the project original budget and schedule had to be held. As the project took off, a completely new design for the accelerator had to be invented, reviewed by outside experts, and fit into a new construction plan that conformed to the original plan. This is where another of Grunder's talents was on full display. This accelerator was being built in a place (southeast Virginia) where there was no existing expertise in the science and engineering of building a high-technology device such as an accelerator. From the time Grunder started his tenure as lab director, he had been able to recruit top talent from existing accelerator laboratories worldwide and hence supply the needed expertise. The recruiting was successful because of Grunder's infectious personality that convinced the candidate not only of the prospect of building something unique from the ground up but also of the excitement of working for Grunder, who seemed—and was—indefatigable. He recruited top-level accelerator designers from Lawrence Berkeley Lab and Fermilab, nuclear physicists who would design the experimental stations from top European labs, and specialized technical talent from wherever he could find it.

He knew the importance of a strong scientific case for any major scientific project, so he successfully recruited two of the day's best-known nuclear physicists. Dirk Walecka from Stanford University became the laboratory's first scientific director [2]. He was followed by Nathan Isgur [3], who was one of Richard Feynman's last graduate students, and was persuaded to leave his position at the University of Toronto. Isgur continued the development of the scientific case for the original project and also its subsequent upgrade, completed in 2018 [4].

The CEBAF project and its surrounding laboratory were being built in Newport News, Virginia, which did actually have an abundance of technical support labor, given that the city also housed the largest private shipbuilding company in the U.S.–Newport News Shipbuilding. Grunder arranged for welders who were recruited from the area to be recertified on the precise welding needed to pipe super cold (2 K) liquid helium to cool the superconducting accelerator structures. He arranged additional technical training courses to be added to the curriculum of the local community college that helped supply technicians to build and eventually operate the accelerator complex.

The CEBAF project was more than a machine. The project needed a surrounding laboratory to ensure that the machine was properly built, installed, and operated for an international scientific user community that eventually reached over a thousand scientists. Grunder, Waleka, and Isgur organized cooperative programs with all of the universities that were within commuting distance to the laboratory so that their proximity could help strengthen undergraduate and graduate science programs.

I consider myself very fortunate to have crossed paths with Hermann Grunder. In 1990, he recruited me from Princeton University to be the laboratory's first Chief Technology Officer. In my first few weeks at the lab, I was introduced with Grunder's entrée to many of the leaders from the city, region, and state who could help strengthen the growth and evolution of the laboratory. It was a whirlwind tour of city councils, legislators at the State and Federal level, and CEO's of regional industries. Grunder showed me by example the benefits of getting to know local and national decision-makers before you might need to ask them for support. During my first year at the laboratory, my time was divided among getting to know both the staff and the technology of the laboratory as it was reaching a crescendo of design activities and beginning the construction of the accelerator complex.

My primary responsibility was starting what was called a "technology trans-fer" program. Key administrators that arranged for funding the lab's science program, such as the Department of Energy which was paying most of the bill for CEBAF, realized that despite significant federal expenditures on public research projects, very little technology that was being developed was trans-ferred to the private sector for potential commercial ventures. This changed with the passage by the U.S. Congress of the Bayh-Doyle Act in 1980, when policies were put in place to enable both national labs and universities to assist with commercialization by patenting and licensing developed technologies. By late spring 1991, I had the rudiments of a technology transfer program in place with the full support of local and regional business groups, and of course Grunder's coaching.

My laboratory responsibilities and my steep learning curve changed dras-tically in early June 1991. I received a call over the weekend from Grunder telling me that, as of Monday morning, I was now responsible for completing the design, construction, and installation of the superconducting accelerator modules. It was an auspicious moment for the laboratory. The accelerator tunnel was ready to receive the first accelerator modules, but only two of the required 42 modules had been made three years after the formal project start, and they had problems. I had six months at most to solve the problems and no more than two years to deliver the full accelerator. If I was going to help "sell" the lab's unique technology to the rest of the world, I now found myself in the thick of getting to know it from the inside out. Fortunately, the group of physicists, engineers, and technicians that Grunder had assembled was very talented. With a fresh face as ringleader and a different take on organizing the accelerator group, we successfully completed the entire accelerator ahead of schedule and slightly under the advertised budget. And most importantly for the future of the lab and other applications of this technology, the supercon-ducting structures exceeded their design goals by a factor of two. Grunder's risky bet of switching the whole project to this new technology had paid off. CEBAF was dedicated as a new laboratory in 1996—with a fortunate name change to Jefferson Lab, and it has since gone on to be one of the most important scientific facilities in the world for basic nuclear physics studies.

The SRF technology developed at the lab enabled the original accelerator to be upgraded to three times its original design energy in this last decade and was also spun-off for use in high-power laser projects (first at Jefferson Lab) and then worldwide. It is a fact of life for modern science that the scale of frontier research projects requires large scientific teams and multi-million-dollar expenditures. However, "one man can still make a difference" with the right combination of skills and charisma [5, 6].

References

1. A series of papers from historian Catherine Westfall provide the detailed references on the history of the CEBAF project and the development of Jefferson Lab. See: a. Westfall, Catherine. *"The Founding of CEBAF, 1979 to 1987"*. 1994. JLS.009, Box 1, Folder 12. The Papers of Catherine Westfall, Jefferson Lab Archives. b. Westfall, Catherine. *"Engineered for Success: Hermann Grunder and the Building of Jefferson Laboratory"*. 2000. JLS.009. Box 1, Folder 16. The Papers of Catherine Westfall, Jefferson Lab Archives, Thomas Jefferson National Accelerator Facility, 12000 Jefferson Ave., Newport News, VA, 23606. c. Westfall, Catherine. *"Jefferson Lab's 1985 Switch to Superconducting Accelerator Technology"*. 1996. Jefferson Lab Information Resources, June 1996. https://misportal.jlab.org/ul/publications/view_pub.cfm?pub_id=11131&CFID=65757&CFTOKEN=ad4d3ba8be27 2dc1-B0874EFE-5056-9E00-FEE4D11299B65369
2. J. Dirk Walecka (b. 1932) is Emeritus Governor's Distinguished CEBAF Professor of Physics at the College of William and Mary. He was recruited from Stanford University in 1986 to be Jefferson Lab's first Scientific Director from 1986–1992. He is a prolific author of many graduate-level textbooks of modern physics. His biographical information is available from AIP's Physics History Network; see: https://history.aip.org/phn/11610034.html
3. Nathan Isgur (1947–2001), was the head of Jefferson Lab's Theory Group and first Chief Scientist of Jefferson Lab, serving in those positions from 1990 until his untimely death in 2001. He was widely known in the theoretical physics community for the development of the 3-quark model of protons and neutrons. He led the scientific program for the initial operation of Jefferson Lab's 4 billion volt electron accelerator and initiated the scientific community's case for upgrading the accelerator to 12 billion volts, a project completed in 2018. Isgur's biography and papers are archived at Jefferson Lab: https://www.jlab.org/ir/archives/findingaids/nisgur.html
4. Thomas Jefferson National Accelerator Facility (Jefferson Lab), 12 GeV CEBAF Upgrade Project, see https://www.jlab.org/ir/archives/

5. John McTague, former head of research at Ford Motor Company and science advisor to U.S. president, Ronald Reagan; quote about Herman Grunder, the first director of the CEBAF project and Jefferson Lab in Newport News, VA, as cited in Ref. 6

6. I. Goodwin, "*CEBAF wins praise for design, but its future is uncertain*", *Physics Today* **39**, 2, 51 (1986): https://physicstoday.scitation.org/doi/abs/10.1063/1.2814883

26

A Bright Light in Virginia

I consider myself fortunate to have book-ended my career as an experimental physicist by working on a great tool of science—the laser. This device, which sprang to life in 1960, spent very little time as a new gadget looking for an application. In the intervening years since its inception, it became an indispensable scientific research tool and ubiquitous in our lives. Worldwide communications and commercial transactions, environmental monitoring, manufacturing, and medicine are all dependent on the use of lasers in their many forms. That innovation continues, extending from the first practical laser demonstrated by Ted Maiman at Hughes Corporation in 1960 [1], which produced short, bright, and tightly focused pulses of red light with a tabletop device, to lasers spanning the spectrum from infrared through X-rays and spanning sizes from the microscopic for microelectronics to room-sized devices for frontier science and heavy manufacturing.

My fascination with lasers began as a 13-year old in 1962, cobbling together enough materials and equipment from very generous mentors to build my first laser [2]. That experience convinced me to study and work hard enough to become and spend my life as a physicist. Despite my fascination with lasers, I spent only the summer after my freshman year in college working directly on them. By then, I had become exposed to the fact that physicists like to say they are interested in everything—from the inner workings of the atomic nucleus to the birth and evolution of the Universe. I was not immune to this grand vision of physics and succumbed. I spent most of the next decade earning my calling card for the trade, the PhD degree, by learning from several remarkable MIT teachers [3]. I then embarked on

© Springer Nature Switzerland AG 2020
H. Frederick. Dylla, *Scientific Journeys*,
https://doi.org/10.1007/978-3-030-55800-0_26

my first extended career stop, joining a research team studying high temperature plasmas at Princeton, with the goal of developing fusion power for future generations [4]. I enjoyed my career at Princeton immensely, but after 15 years, lasers were calling out to me again.

In 1990, I was enticed by another remarkable and visionary physicist, Hermann Grunder (described in Chap. 25) to join his team in building a new electron accelerator laboratory in Newport News, Virginia [5]. Hermann was recruited by a group of Southeastern U.S. universities who won a nationwide competition to build the new federally funded nuclear physics research facility. This group realized an experienced machine builder and laboratory director would be required to build the more than $200 (USD) million project. Hermann convinced me to leave my comfortable career at Princeton and help him build the new accelerator. But he had an additional enticement for me. He wanted me to figure out how to build (and pay for) a unique and very powerful laser that could be added to the accelerator site as a complementary project.

I accepted Herman's call and in the fall of 1990 I established a new home in Virginia and began what turned out to be another major chapter in my life as a physicist—and this time lasers were back on my scientific agenda. The primary purpose of this new national laboratory (now named Jefferson Lab, in honor of Virginia's extensively science-minded patriot and native son) was to build and operate a large (5 km) and powerful (4 billion volt) electron accelerator that would be a new microscope for understanding the structure within fundamental particles that make up the atomic nucleus, namely, protons and neutrons.

An essential part of my training for my new Jefferson Lab job was to manage the group building the accelerator—the first large scale use of a new type of superconducting device that would be efficient to operate and straightforward to upgrade as the facility aged. Hermann sensed that if this "training" went well, I would be in good stead to help build a second project at the laboratory—a powerful and versatile new kind of laser that would use the same kind of powerful electron beam that intrigued the nuclear physicists who founded Jefferson Lab.

This new type of laser was called a free electron laser or FEL [6]—a name coined in 1976 by its inventor John Madey (then at Stanford University). All other types of lasers generate powerful light beams by exciting electrons trapped within a solid, liquid, or gaseous medium. Generally, the particular color of light emitted by a laser is limited by the choice of the laser medium. The extracted light power is very definitely limited by the ability to "pump" input energy into the medium and extract waste energy in the form of heat that is not converted to laser light. A free electron laser operates on electrons freely beaming down the evacuated space of an accelerator tube, thus unimpeded by the constraints of a lasing medium. The big problem is that the required powerful electron beam is produced by a large and expensive accelerator—so this is not "free light". However, by using an efficient super-conducting accelerator and extracting a lot of laser light, the cost of producing a watt of laser light per watt of input electricity could, in principle, be made very attractive. A second very attractive advantage of making laser light with an FEL was the ability to change the color or wavelength of light. Stated simply, the goal of this new laser project at Jefferson Lab was to make a lot of laser light very cheaply over a wide wavelength range. As with most first-of-a-kind projects, easier said than done.

Fig. 26.1 The author standing behind the key structure of a free electron laser—the "wiggler"—an array of magnets that extracts powerful laser light from a traversing electron beam. *Credit* Jefferson Lab

The story at this point in the project timeline (1991) becomes a tale of how a group of scientists and engineers managed to assemble a critical mass of support to launch a new research device, and in particular, owing to its size and complexity, one that required an investment of the order of 10–20 million dollars. If a planned device has short-term practical applications, the proposers might manage to secure industrial or other private funding. If it is largely a research tool for basic science, the funding source is public and such projects are constantly being proposed by bright scientific groups with expected and significant competition. Since we believed this powerful laser would find applications for both basic and applied research, we developed a hybrid approach and formed a group of potential users from both the university and industrial research communities to build a case for the machine, generate a portfolio of initial uses, and help solicit the required start-up funds, or obtain donations. Our modest group of scientists and engineers produced our first proposal in mid-1991, describing a unique stand-alone machine that could generate more than a kilowatt of laser light—widely tunable across the light spectrum from the infrared to the ultraviolet. This first proposal was met with skepticism by prior free electron laser builders because it would provide 1000 times more light than had ever been generated by a FEL—and prior devices had only provided infrared light. We intended to broadcast the advantages of this much more powerful laser with expanded tunability. When any group of scientists propose a novel (and expensive) new machine, it is important from the start to have potential users of the device brag about its capabilities rather than the builders—make the case customer-driven, so to speak.

This is not an easy task when one is promoting something new and yet to be built, especially when success in obtaining the funding to deliver the project remains uncertain. We spent the next three years building a collaboration of users to present a strong case for the machine. On the industrial side we approached seven large companies that still maintained significant corporate research laboratories for their interest and financial support. In addition, we approached a group of largely local universities who could both support the industrial applications and propose basic research problems of their own. By 1994, we submitted a proposal to a funding opportunity advertised by two U.S. funding agencies (the Department of Energy, the major sponsor of Jefferson Lab, and the Department of Commerce, which oversees research support for industrial needs). Our proposal requested $25 million to build this powerful laser and a surrounding laboratory that would allow 10–20 user groups to set up their experiments or processing equipment for use of this heretofore unavailable range of laser light. As a sign of the industrial support, this cohort pledged a donation of more than $10 million of equipment for initial outfitting of the user laboratories. In addition, a pledge from the state of Virginia for $3 million would finance the building housing the laser and the user laboratories. Therefore, the requested federal government funds were needed for only half the project cost—an attractive cost-sharing model. However, despite our collaboration's enthusiasm for the proposal and their significant donation of time to develop and present the proposal, the pot of government funding we aimed at in 1994 was too small and the competition was fierce. Our proposal received excellent marks from the teams the agencies hired to review and rank the proposals, but the proposal didn't make the cut—only 0.1% of the submitted proposals were funded. When one asks for significant public funds, the journey is not easy. Our proposal was refined and the group tried again to unlock the coffers.

Jefferson Lab is conveniently located next to a much larger national laboratory operated by NASA—the Langley Research Center for Aeronautics and Astronautics. We brought them in as a partner and arranged for a joint review by a panel constituted by both the Department of Energy and NASA. The review was held in the spring of 1995, led by our now seasoned FEL industrial and university scientific collaborators. Everyone participating felt good about the review afterwards and all were hopeful for a path forward to start the project. By summer's end, a short letter from Hazel O'Leary, the Secretary of the Department of Energy, arrived. The key sentence in the letter praised the proposal team for an innovative proposal that was a "model for scientific research public-private partnerships" and ended with "good luck in finding the money...".

Given the success rate in securing the funding to get this project going, we expected that some of our partners might give up on us. But they didn't, and we immediately explored another avenue for the missing half of the needed funds to start building the laser. Two of our industrial partners helped us get a foot in the door with the U.S. Office of Naval Research—the nation's oldest scientific funding agency, founded immediately after WWII. By late 1995, $10 million of Navy funds was finally secured, and with some loaned excess parts from Jefferson Lab's existing accelerator program, the team started to build and deliver the world's most powerful, tunable laser to our patient user community. A vagary of U.S. national politics that year would make the construction project a cliffhanger. The Navy funds for the project were included in the federal government's fiscal year 1996 appropriation. For the early part of this year, the Republican led U.S. Congress was in a stalemate with the Democratic U.S. president, Bill Clinton and the federal government was shut down for all but essential services for two extended periods early in the funding period. This was long enough to delay release of the Navy funds required to start the project until May 1996. To add to the drama, the Navy funds came with a clock. The funds had to be spent within two years of the

Congressional appropriation, but since the funds were almost 6 months late in release, the entire project had to be compressed into 18 months, along with 14 different reviews of the project during this same period by our government overseers, to make sure public funds were being wisely spent. Sounds like a pressure-cooker? With the help of an extraordinary team of colleagues, the construction project was completed on September 30, 1997, the last day of the 18-month schedule, and the careful process of commissioning a new machine began. These devices are complicated in their first incarnation. Turning it on and coaxing light out the first time is a deliberate step-by-step process. In March 1998, it produced its first laser light, eclipsing the power record of all previous FELs in the world, and by the summer of 1999, the goal of producing over a kilowatt of laser light was achieved [7].

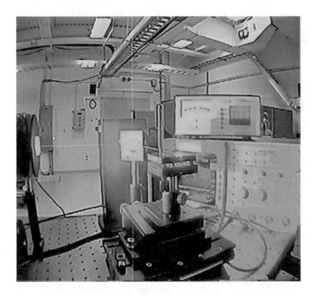

Fig. 26.2 Powerful green laser light illuminating one of the Jefferson Lab experimental labs. *Credit* Jefferson Lab

Over the next four years, the now very patient user community got its first chance to use this new light source. Experiments spanned a wide range from chemistry, biology, materials science, materials processing, and biomedicine. (Interested readers can explore these experiments, described in Ref. [8].) While these user experiments were being done, our laser building team laid the groundwork to make the laser ten times more powerful, with up to 10 kW of light output. We applied to the Office of Naval Research for this upgrade of the initial laser, and we diversified our funding base by successfully bringing on the two additional funding agencies to extend the laser wavelength to the longer wavelength far infrared and to shorter wavelengths in the ultraviolet. By 2004, the upgraded laser was on the air and produced a record power of 14 kW of infrared laser light [9]. This remains the record for the most powerful tunable laser ever built.

One of most successful user experiments was done by our partners at Harvard's Massachusetts General Hospital, who showed how the powerful and tunable infrared light could be used to destroy fat cells without harming neighboring non-fat tissues [10]. The process has since been put into clinical practice using lasers custom built for clinical use. With such an interesting start, we expanded our consortium to include three other medical schools and attempted to fund an expansion of the user facilities for additional biomedical applications. Our first attempt at funding was not successful, which is typical for scientific funding requests, as this story relates. But this is where my connection to this project ends. In late 2006, I was asked to head up the American Institute of Physics in College Park, MD. Despite my ties to this unique laser, I could not turn down this offer and my career with lasers ended. But what a ride it was.

My colleagues at Jefferson Lab continued extending the capabilities of the FEL to the ultraviolet [11]. Unfortunately for the team, which had devoted nearly 20 years to the project, it was shut down in 2011. Ironically, their successful efforts rejuvenated this unique type of laser and over the last decade their success helped to initiate nearly a billion dollars of investment in the construction of the first X-ray laser experimental facilities in the world— one in Stanford, CA, and one in Hamburg, Germany [12]. Biochemists and biomedical scientists have long dreamt of extending the magic of lasers to the regime of X-rays. With sufficiently bright and ultra-short pulses of X-rays, the structures of the essential building blocks of life such as proteins could be decoded much more quickly than with conventional X-ray sources. That dream is coming to fruition thanks to free electron lasers.

References

1. Theodore H. Maiman, *The Laser Inventor*. Springer International Publishing, Cham, SW (2018)
2. See Chapter 16; an abridged version of this story was previously published in *Physics Today Online*, 12 August 2010
3. See Chapters 17 and 27 for my encounters with two remarkable MIT professors
4. See Chapter 18 and a useful general reference on the first 50 years of fusion research, including the program at Princeton University: Dale Meade, *Nuclear Fusion* **50** (2010) 014004; https://doi.org/10.1088/0029-5515/50/1/014004

5. Christoph W. Leemann, David R. Douglas, and Geoffrey Krafft, "*The Continuous Electron Beam Accelerator Facility: CEBAF at the Jefferson Laboratory*", Annual Reviews of Nuclear and Particle Sciences **51**, 413–450 (2000). https://doi.org/10.1146/annurev.nucl.51.101701.132327

6. Charles A. Brau, '*Free Electron Lasers*', Science **239** (4844) 1115–1121(1988); https://doi.org/10.1126/science.239.4844.1115

7. G.R. Neil et al, *Physical Review Letters* **84** (4), 662 (2000)

8. H. Frederick Dylla and Steven T. Corneliussen, "*Free-Electron Lasers Come of Age*", *Photonics Spectra*, p. 65, August 2005

9. G.R. Neil et al, *Nuclear Instruments and Methods in Physics Research, Sect. A* **557** (1), 9–15 (2006)

10. R. Rox Anderson et al, *Lasers in Surgery and Medicine* **38**, 10, 913–919 (2006), https://doi.org/10.1002/lsm.20393; Fernanda H. Sakamoto et al, *Lasers in Surgery and Medicine* **44**, 175–183 (2012); https://doi.org/10.1002/lsm.21132

11. S.V. Benson et al, *Nuclear Instruments and Methods in Physics Research*, Sect. A 649 (1), 9–11 (2011); https://doi.org/10.1016/j.nima.2010.12.093

12. Pellegrini, C. "The history of X-ray free-electron lasers", *European Physical Journal H* **37**, 659–708 (2012). https://doi.org/10.1140/epjh/e2012-20064-5

27

Gravity's Songs

Early one September morning in 2015, the last of four predictions Einstein had made when he completed his general theory of relativity [1] a century earlier was confirmed using one of the most remarkable scientific instruments ever built. Rai Weiss, now an emeritus professor of physics at MIT, first described the possibility in an informal laboratory report in 1972—how one might detect gravitational waves using intersecting laser beams reflecting from carefully isolated mirrors to detect incredibly small changes in the positions of the mirrors. I was just beginning my graduate studies in physics at MIT, working in the same laboratory where Rai had done his graduate research—Professor Jerrold Zacharias's Molecular Beam Laboratory. Rai finished his degree a decade before me and was launching his career as a professor and researcher. He excels in both professions, which he continues at the time of writing, well into his ninth decade.

There is no more important decision for a scientist beginning a graduate education than the choice of a thesis advisor. For a few, this choice may only be an institutional necessity for the required signature of a senior faculty member when the penultimate written thesis is reviewed, signed, and deposited in the library archives. For most students, including the present writer, the thesis advisor is much more—a coach, a teacher, a mentor, a critic, and often a lifelong friend and colleague. I was fortunate to have two advisors. Professor John G. King, who took over directing Zacharias's lab, was my official advisor. I describe my interactions with this unique educator and physicist in Chap. 17. Although I was not one of Rai's official graduate students, my education as a physicist was immeasurably impacted by my frequent interactions with Rai and his other students. He had a curiosity

© Springer Nature Switzerland AG 2020
H. Frederick. Dylla, *Scientific Journeys*,
https://doi.org/10.1007/978-3-030-55800-0_27

about physical measurements that was infectious. Could a measurement be made that would tell us more about how the world worked—from the inner workings of the subatomic world to the clockwork of the Universe? Usually the answer to such a sweeping question was affirmative in many fields, but to move the query to a working instrument that pushed the measurement to frontier physics meant understanding how to extract microscopic needles from mountainous haystacks. During my MIT student days, I saw Rai and his students work on projects that spanned both ends of the yardstick of physical measurements. For his graduate research, he invented a detector for single atoms that is still unsurpassed in sensitivity. For my Ph.D. thesis work, I needed such a detector and tried to improve on Rai's gadget, but I eventually used his design. During my graduate career, Rai and his students undertook difficult problems such as stabilizing laser light and improving the sensitivity of infrared detectors. Both activities would be incorporated in subsequent projects across the scientific world that led to future Nobel prizes in physics.

Professors at research universities such as MIT have teaching responsibilities in addition to their research activities. After preparing for a course on Einstein's theory of relativity, Rai began thinking about whether a laser interferometer could be used to detect gravitational waves—the rippling of the fabric of space due to the motion of any mass in the Universe. Such waves can be visualized as analogous to the ripples that are seen on the surface of a pond when a pebble is thrown in. As the water waves travel outward from the point of impact, the amplitude of the waves decreases from their initial height. Einstein's 1916 tour de force, his general theory of relativity, describes the motions of the Universe just as we observe and measure its structure and evolution. As radical as the theory was at its inception, the first of four key predictions was confirmed with a famous measurement during a 1919 solar eclipse of the deflection of starlight due to the Sun's gravity [2]. A century later we are all dependent on general relativity's minute corrections to the measurements of the microwave signal transit time between our mobile telephones and orbiting satellites that comprise the Global Positioning System. The last of the four major predictions from the theory to be confirmed was the detection of gravitational waves. The theory predicts that the motion of any body in the Universe would be accompanied by waves rippling across the fabric of space at the speed of light. The amplitude of such waves, which is a measure of the localized stretching and compaction of space as a gravitational wave traverses it, is directly calculable from the theory. So why did it take a century between Einstein's prediction of the existence of gravitational waves and their detection? Except in the extreme case where we, the observers, are in the neighborhood of a catastrophic event in the Universe (where we would not survive to commemorate the event), the amplitudes of gravitational waves are truly minuscule, resulting in infinitesimally small displacements of any reasonably sized detector.

A series of attempts to detect gravitational waves were made in the 1970s using massive metal bars that were ingeniously suspended and isolated from ground motions [3]. As a gravitational wave emanating from a raucous celestial event passed through the bar, attempts were made to detect the microscopic shaking of the bar, with net motions much smaller than atomic dimensions, amid the cacophony of interfering signals. These experiments were not successful, and for the next few decades they cast a pall on experimental gravity research. Rai was undeterred by this pall. In the midst of these early attempts with large bar detectors, Rai's 1972 MIT lab report presented a detailed calculation of how a laser interferometer could be used to detect gravitational waves—provided the device was large enough to boost the estimated tiny signals. A unique feature of his report was its comprehensiveness. Nearly every experimental measurement that bounds or extends the frontiers of our knowledge of physics is a battle to find and extract a signal from an armada of competing noise sources. Even the non-scientist may be familiar with this task from dealing with weak analog television or radio signals in the broadcast era, or with the first generation of mobile telephones.

What Rai described in his 1972 report was a complete analysis of the signal strength and competing noise sources in order to predict the sensitivity of a large laser interferometer for detecting a passing gravitational wave. Rai never published this report in a scientific journal, even though it has now obtained legendary status as one of the most important "non-publications" in the still young field of gravitational astronomy. He has stated emphatically that the report described an idea—a very well elaborated idea. But he felt strongly that formal scientific publications warranted the full story from the expression of the idea to its fruition. A long journey took this idea to a fully developed credible design that survived multiple peer reviews, a tortuous battle in the U.S. Congress to appropriate nearly $200 (USD) million (more than the U.S. National Science Foundation had ever dedicated to a single project), and by the turn of the century, successful construction of the world's first major gravitational wave "telescope".

In 2005, the two facilities that comprise the Laser Interferometer Gravitational Observatory (LIGO) achieved their design sensitivity [4]. The facilities were identical, two perpendicular, large diameter (~1 m) vacuum pipes, 4 km in length. The facilities are located more than 3000 km apart, one in the central Louisiana pine forests and one in the high eastern plains of the state of

Washington. Jointly operating two nearly identical instruments allows uncorrelated ground noise and anthropogenic noise sources to be discounted in favor of the weak extraterrestrial gravitational wave signals.

Fig. 27.1 LIGO detector site near Hanford, Washington. *Credit* Courtesy Caltech/MIT/LIGO Laboratory

Despite the fact that, when these devices became operational, they were able to make the most sensitive length measurements ever imagined, no member of the LIGO scientific team, particularly the key Caltech scientist, Kip Thorne, who shares responsibility with Rai for the birth and eventual success of LIGO, ever felt that these first instruments would be sensitive enough to detect gravitational waves. Kip Thorne is a revered theoretical physicist, a student of the late Princeton professor John Wheeler who encouraged physicists to keep working on Einstein's general theory and coined the name "black hole" for the end state of certain stars. In a fortunate coincidence of space and time, Kip Thorne and Rai were both attending a 1975 NASA meeting in Washington, DC. During a long night's discussion of potential topics for launching a new experimental physics research group at Caltech, the focus became gravitational wave research, including Rai's ideas for a long baseline interferometer. Thus, began a long collaboration that brought in Caltech as the lead partner with MIT for the design and subsequent operation of LIGO. Thorne and his students provided the initial estimates for the possible sources of gravitational waves that might be strong enough to be detected by LIGO. Catastrophic events such as two colliding black holes were deemed to be the most detectable events. But by the time LIGO first became operational, the sensitivity of LIGO was deemed insufficient to detect such events. To cloud the prospects of a new facility even further, there were no reliable estimates of how many black holes were within observational range and how frequently two such objects might collide. This was quite an unusual situation for an expensive scientific instrument that would take decades to design, fund, and build [5].

The subsequent history of LIGO is testament to both Rai and Kip Thorne's scientific tenacity and the hard work of the small LIGO staff (<100) and the growing LIGO scientific collaboration (~800). When the first generation LIGO was operated between 2005 and 2010, improvements in sensitivity were made with incremental changes to the hardware and signal processing. Both LIGO facilities were shut down in 2010 to allow a significant upgrade in the laser hardware. In September of 2015, as the upgraded machines were going through final testing, one of the most important observations in modern science was made.

On September 14th 2015, at 09:50 GMT, a signal was detected in the Louisiana facility, and 20 ms later the same signal was detected in the Washington facility. This signal, lasting a mere 200 ms, had all the predicted signatures of two colliding black holes. Given the importance of this finding, the LIGO team took until February 2016 to announce the discovery, along with a simultaneous publication detailing the specifics [6]. Given the sophistication of the calculations that were now possible based on Thorne's initial work, and the quantitative measurement of the signal strength from the LIGO detectors, the measured signals had the predicted signature of the last four orbits of a death spiral of two colliding black holes, each about 30 times the mass of our Sun, in an event that occurred 1.4 billion years ago. Because the signal was detected by both LIGO's instruments, the death spiral could be approximately localized to within a circular segment of the cosmos.

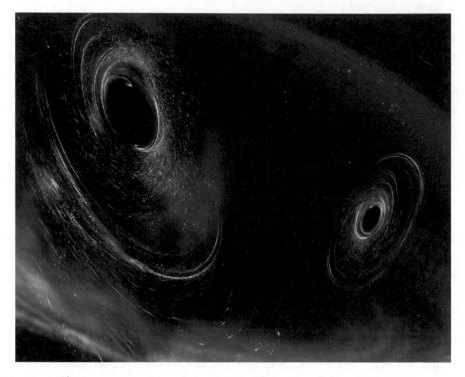

Fig. 27.2 Artist's conception of the spiral dance of colliding black holes. *Credit* Aurore Simonnet, LIGO/Caltech/MIT/Sonoma State

Throughout his career-long dedication to LIGO, Rai had hoped that a firm detection of gravitational waves would be made before the centenary of Einstein's presentation of his general theory. Well, this remarkable scientist got his wish—Einstein first presented his general theory to the Prussian Academy of Science on November 25th, 1915.

The field of gravitational astronomy has exploded since the first detection of gravitational waves in September 2015. Numerous other colliding black hole collisions have been detected, and these events now serve as benchmarks to further improve the sensitivity of the interferometers. In August 2017, the most remarkable discovery to date was recorded—a collision of two neutron stars [7]. Because one of LIGO's sister machines, the Virgo instrument in Italy, was also collecting data, the trio of gravitational wave detectors could localize to a reasonable degree where this celestial event was occurring in space. The observation was broadcast to an international array of telescopes that monitor the cosmos in all wavelengths, from the radio range through visible light and short wavelength X-rays to gamma rays. The event became the first to monitored in the new field of "multi-messenger" astronomy, allowing information heretofore unavailable. One unique observation was the detection of a gamma ray burst nearly coincident with the detection of gravitational waves emanating from the collision. This observation confirmed that gravity waves travel at the speed of light, as Einstein predicted. We also learned from this event the origin of certain elements heavier than iron in the periodic table. We know that they are born in the primal furnace of the last moments of the collision, because the signature of their birth was broadcast to this array of primed telescopes.

Fig. 27.3 Professor Rainer (Rai) Weiss at the New York Science Festival, 2010. *Credit* Matt Weber

Rai, Kip Thorne, and Barry Barrish, the Caltech physicist responsible for building the LIGO scientific collaboration, shared the 2017 Nobel Prize in Physics. As Rai is approaching his 10th decade, he felt the prize was a distraction from his continuing work on LIGO. He did enjoy the ceremony though, using his prize money to bring a large group of the LIGO staff with him to Stockholm. Since the prize, he has been concentrating on helping LIGO collaborators design a next generation LIGO with ten times the length (40 km arms) and ten times the sensitivity. He knows he can find a way to make sure it doesn't cost ten times as much.

References

1. Albert Einstein, *Relativity: The Special and General Theory*, was first published by Einstein in 1916, shortly after his release of the General Theory. It was written for a general scientific audience and is considered a unique view of a great scientist's exposition of his own work. The work was translated for its first English edition in 1920 and has appeared in many editions hence, see for example: *Relativity: The Special and General Theory*, https://www.bartleby.com/173/

2. The original measurement by Eddington's team is published here: Dyson, F. W.; Eddington, A. S.; Davidson, C. (1920). *"A Determination of the Deflection of Light by the Sun's Gravitational Field, from Observations Made at the Total Eclipse of May 29, 1919"*. *Philosophical Transactions of the Royal Society A: Mathematical, Physical and Engineering Sciences*. **220** (571–581): 291–333. https://doi.org/10.1098/rsta.1920.0009. For a general reader account of this famous measurement and the aftermath, see: Daniel J. Kennefick, *No Shadow of a Doubt: The 1919 Eclipse That Confirmed Einstein's Theory of Relativity* (2019), Princeton University Press, Princeton, NJ (2019)

3. For a poignant account of Joseph Weber's career long and unsuccessful search for gravitational waves see the following memoir by his lead engineer, Darrell J. Gretz, *Early History of Gravitational Wave Astronomy: The Weber Bar Antenna Development* in the *APS Forum of History of Physics Newsletter*, Spring 2018, American Physical Society, College Park, MD; https://www.aps.org/units/fhp/newsletters/spring2018/weber.cfm

4. The website of the LIGO Scientific Collaboration: www.ligo.org

5. Janna Levin, *Black Hole Blues and Other Songs from Outer Space*, Alfred A. Knopf, New York (1916)

6. B. P. Abbott *et al.*, *Observation of Gravitational Waves from a Binary Black Hole Merger*. Phys. Rev. Lett. **116**, 061102

7. B.P. Abbott et al., *Multi-messenger Observations of a Binary Neutron Star Merger*. Ap. J. Lett. 848:2 (2017)

Part III
Science Policy Matters

28

Innovation in a Young Nation

Historians and economists often consider that the spark of America's economic engines was ignited during the second half of the twentieth century. Prodigious inventions and discoveries resulted from the post-World War II economic boom and fears of second-class world status in technology after the Soviet Union's successful Sputnik launch. The entire world economy is still benefiting from the innovation which poured from the corporate research laboratories of America's blue-chip companies and the fast-growing populations of America's research universities.

However, author Charles Morris expertly illustrates how the tradition of innovation in America began just as the new country was getting started. Morris' 2012 book, *The Dawn of Innovation* [1] describes a young nation's transformation from a minor player on the world stage at the opening of the nineteenth century to eclipsing the British Empire at the end of the century as the world's largest economy and manufacturer. Some of the more famous inventions marking America's developing industrialization include Robert Fulton's steamboat (1777), Eli Whitney's cotton gin (1793), and Samuel Colt's systemized manufacture of firearms with interchangeable parts (1836).

Morris, a distinguished banker and lawyer by trade, has authored more than a dozen books on American history, with an emphasis on events and personalities having a profound influence on America's economic development. With the subtitle *"The First American Industrial Revolution,"* his book

This article was adapted from a book review of Morris's book by H. Frederick Dylla that was previously published in *Physics Today* **67**, 5, 51 (2014).

© Springer Nature Switzerland AG 2020
H. Frederick. Dylla, *Scientific Journeys*,
https://doi.org/10.1007/978-3-030-55800-0_28

provides an engaging account of the remarkable technologies, businesses, and distribution systems that were put in place across the American continent as settlements rapidly moved west either on foot or by wagon, boat, and rail.

Fig. 28.1 The state opening of the Great Exhibition in London, England, in 1851. *Credit* Color lithograph by Louis Haghe, from the collection of the Victoria and Albert Museum, London

Some key numbers tell the story. In 1800, Britain was the leading global provider of raw materials, such as coal and iron, as well as finished products, with America's output a mere one-sixth of that of the British territories. At the 1851 Great Exhibition of the Works of Industry of all Nations [2] held in London's Crystal Palace, Victorian England continued to celebrate its mastery of the world's economy—but the United States gained attention for its ability to mass produce interchangeable parts. Moreover, we were catching up; U.S. production had grown to approximately 30% of the British output. In the 1880s, U.S. output had achieved parity, and by the outbreak of World War I, it exceeded British output by a factor of 2.3.

This catch-up century began with war preparations—a familiar boost to economic development. Morris describes the naval arms race during the War of 1812 (1812–1815), in preparation for a major battle between the United States and Britain in the Great Lakes region. Because of its strategic location separating the U.S. and Canada, both sides anticipated major military engagements on Lake Ontario or Lake Erie. In 1812, the U.S. Navy had one ship supporting 16 guns and the British had six ships supporting 40 guns. By the end of this short war, the respective firepower was essentially equal, and more than 100 ships were armed and manned on both side of the lakes. Fortunately for the sailors in both fleets, there was never a major battle. (Note the uncanny similarity to the Cold War arms race between the United States and the Soviet Union that thankfully never culminated in attack.) The central reason a significant battle never came to pass appears to be an asymmetry in the fleets' capabilities. The British were superior in sailing and maneuvering in stormy seas. The American fleet performed best in calm seas and it had amassed a larger number of long-range cannons, which favored longer distance engagements. This disparity meant that no time was optimal for a good fight. The investment nevertheless proved profitable for America; the necessity to rapidly build up its fleet jump-started the shipbuilding and attendant industries, such as lumber and iron.

Morris traces American nineteenth century industrial development through multiple pathways. Power generation sprang up on almost every stream east and west of the Appalachians. Water turbines were developed for major dams, and they are still in use today for generating hydroelectric power. Shipping and transportation networks replaced horse and carriage with canals and steam ships. A rail system crisscrossed the nation by mid-century and quickly grew to an established railroad network, also still used today.

At the dawn of the nineteenth century, the United States was primarily a supplier of raw materials to Britain and the continent; it had to import most finished products. However, U.S. industry quickly evolved beyond agriculture to basic household necessities such as clothing, furnishings, and chemical products derived from agriculture (lard and whale oil), and then to the heavy industries that fueled (oil) and transported (steam engines, locomotives) the nation, eventually propelling the U.S. economy to surpass Britain's. Along the way, a banking industry was established with an elite financier class, with the likes of Mr. Rockefeller, Carnegie, Morgan, and Vanderbilt, who reaped extraordinary fortunes from this exploding economy.

Much of the early development of these technologies and the underlying inventions, such as the steam engine and the Bessemer furnace for steel, were British born. What fed the American economy to grow so rapidly and overtake the mother country? Morris identifies three major factors at the beginning of the century that led to America's success: (1) near universal male suffrage for the predominant citizenry (except for the non-white population) leading to widespread participation in local and national government; (2) mass public education, fostering the growth of a significant middle class participating in the economy; and (3) a seemingly unlimited availability of natural resources from an untapped continent. Later, in the midst of the country's most trying period, the Civil War, President Lincoln's fractious Congress passed three pieces of legislation in 1862 that strongly underpinned the country's continued economic growth and prosperity as the war ended: the Homestead Act, the Transcontinental Railroad Act, and the Land Grant College Act. Few pieces of federal legislation before or since have had more effect on the American economy, with the exception of the "GI Bill" after World War II which provided free college education to any returning war veteran.

Morris completes his account of this first century of American innovation with what he characterizes as a prologue and epilogue. For the prologue, he recounts how the new American nation was able to reinvent the industrial revolution and outplay Britain to become the world's economic leader. In his epilogue, he contrasts this history with the current landscape, where the United States has taken the role of incumbent, facing a fast-growing China for twenty-first-century dominance. There are lessons unfolding on both sides of this economic race; we must consider the values of participatory democracy, a strong public education system, and respect for natural resources. And in this century, we must find a balance between economic power and global welfare.

References

1. Charles R. Morris, *The Dawn of Innovation: The First American Industrial Revolution*, Public Affairs, New York (2012)
2. Jeffrey A. Auerbach, *The Great Exhibition of 1851: A Nation on Display*, Yale University Press (1999)

29

Big *and* Small Science: No Need to Choose

Funding for "big science" versus "small science" is a contentious topic that often fuels debates among scientists and funders of science. Big science is characterized by major projects such as high-energy particle accelerators and major telescope facilities. These projects are conceived, designed, built, and used by large groups of scientists and engineers. Small science typically involves a principal investigator and a small group of students conducting laboratory research. The scientific enterprise requires both: most scientists are trained in small science, and big science often requires hundreds of small science investigations that are integrated into the grand plan.

My own career in physics is a good example of this co-mingling. As a graduate student, I devoted four years of my graduate student life as principal investigator of a small-scale research project and I would later become involved in some very large-scale scientific programs. Demonstrating the capability to conceive, design, build, implement, and summarize the key results of a scientific investigation is the primary requirement for a Ph.D. in experimental science. In today's frontier research climate, this is usually the last time that any scientist works in the research community as a lone agent—progress requires working with research teams of varying sizes and from various scientific disciplines.

I take as a typical example of this debate a letter by John Waymouth that was published in *Physics Today* [1]. Waymouth laments over the funds that the U.S. federal government has dedicated to large scientific facilities for high-energy physics as opposed to the fraction of funds dedicated to the lower-energy physics and chemistry of practical electronic devices. I have known Waymouth, an R&D pioneer for electric lighting products, since

© Springer Nature Switzerland AG 2020
H. Frederick. Dylla, *Scientific Journeys*,
https://doi.org/10.1007/978-3-030-55800-0_29

my MIT student days and often interacted with him at local and national meetings of scientific societies. As the R&D director at GTE Sylvania's Lighting Group through the late 1980s, Waymouth led valuable electric lighting research efforts that produced usable products and economic gain for his company. The development of practical, efficient, and safe lighting products requires materials science, vacuum science, plasma science, and their associated technologies. Anyone visiting the lighting aisle in a hardware store knows that the Edison light bulb has evolved into a dizzying array of options involving incandescents, fluorescents, and LEDs—and that is just the consumer market.

Would the US be developing more consumer products if we had in the past devoted more funds to research that directly affects our daily lives than to high-energy physics? I believe that the reverse is true.

Following the lighting example, the new generation of LED lighting is made possible by ultrahigh vacuum technology that was paid for by federal investments in hardware needed for high-energy physics, fusion, and space projects. Plasma lighting and plasma displays owe similar debts to the investments made in basic plasma physics by the nation's fusion program [2].

Basic research generates financial returns to the nation's economy through several pathways. It usually takes many years to produce innovative products derived from the original research, but numerous studies have shown this return to be many times the original investment. The developed world's investment in high-energy physics over the last 50 years has been in the tens of billions of dollars. However, the return to the world's economy from just one invention from this field—the World Wide Web—is orders of magnitude higher. The less visible return from basic research is the scientific and technical workforce it creates, a human resource that can, in turn, more effectively advance commercial endeavors through the power of critical thinking and the scientific method.

References

1. John F. Waymouth, "*Investing in electron-volt physics*", *Physics Today* **64**, 9, 10 (2011)
2. H.F. Dylla, "*The Development of ultrahigh and extreme high vacuum technology for physics research*", *Journal of Vacuum Science and Technology*, **A21**, S25 (2003)

30

Reminding All of Us That Basic Research Pays Off

The total cost of all basic research from Archimedes to the present," *my former teacher, the late physicist and international science leader Victor Weisskopf wrote in 1969,* "is less than the value of ten days of the world's present industrial production. [1]

Fig. 30.1 Victor Weisskopf (1908–2002). *Credit* Photograph by Heka Davis, courtesy AIP Emilio Segre Visual Archives, *Physics Today* Collection

© Springer Nature Switzerland AG 2020
H. Frederick. Dylla, *Scientific Journeys*,
https://doi.org/10.1007/978-3-030-55800-0_30

Weisskopf's words are highly relevant half a century later as governments who fund most basic research activities, and even some private concerns, debate their reasons for making such investments.

Few people understood the value of basic research as well as Weisskopf did. His career included contributions to quantum mechanics and the Manhattan Project, a stint as director-general of the European Organization for Nuclear Research (CERN), and chairmanship of MIT's physics department, one of the world's largest. Commenting on the launch of Sputnik during the Cold War, Weisskopf wrote: *"Fundamental research creates the intellectual climate in which our modern civilization flourishes. It pumps the lifeblood of ideas and inventiveness not only into the technological laboratories and factories, but into every cultural activity of our time. The case for generous support for pure and fundamental science is as simple as that."*

In the U.S., support for basic research by the federal government has generally been quite strong, although it has not always translated into the level of funding that many feel is necessary for the United States to sustain a vibrant research enterprise with the resulting benefits to the economy, health, and security of the populace. For example, U.S. federal government funding of research has remained relatively static at only a 0.4% share of the nation's Gross National Product (GNP) for the four decades between 1977 and 2017 [2]. Groups like the Council on Competitiveness, the Business Roundtable, the National Association of Manufacturers, the U.S. Chamber of Commerce, and the National Academy of Sciences have called for restoring research competitiveness in numerous reports—including a seminal report from MIT [3]. That report, *"The Future Postponed – Why Declining Investment in Basic Research Threatens a U.S. Innovation Deficit,"* offers examples of how greater funding would enable advances in treatments for Alzheimer's disease, space exploration, quantum information technologies, energy, photonics, and many other important areas.

Can knowledge gained by basic research conducted out of pure curiosity immediately be put to practical purposes? Consider the case of quantum mechanics—the arcane, highly mathematical, and puzzlingly counterintuitive field that Weisskopf helped advance. Quantum mechanical theory explains the realm of the infinitesimally small. At first quantum mechanics offered only enormously intriguing intellectual payoffs, but it came to underlie the electronics pervading practical modernity—computers, mobile phones, global positioning, musical devices, digital cameras, and the tools of medical diagnosis.

The pattern—basic research leading eventually to technological transformation—is an enduring one, as Weisskopf pointed out in highlighting the return on investment from all basic research since Archimedes. However, the timescale for returns on many worthwhile scientific endeavors can be decades. Radio, television, and radar first required James Clerk Maxwell's nineteenth century studies of electromagnetism. Einstein's seemingly esoteric theory of general relativity published in 1915 is absolutely necessary for the accuracy of our valued GPS location devices. James Watson and Francis Crick could not have predicted the full extent to which their solving the structure of DNA in 1953 would benefit medicine—and life.

Weisskopf also understood the important secondary effects stemming from conducting basic research. Compact superconducting magnets developed for accelerators, for example, enable MRI devices. Likewise, compact, high-resolution light detectors developed for telescopes revolutionized digital photography. Researchers at CERN multiplied the Internet's usefulness enormously by creating the World Wide Web.

After I left the training provided by Victor Weisskopf and other dedicated MIT teachers, I was fortunate to spend more than 30 years of my career working as a physicist at two U.S. Department of Energy national laboratories. At Princeton University's Plasma Physics Laboratory, the main mission was the development of fusion energy—a technology that won't be applied for many decades to come (see Chap. 18). However, as the field worked toward energy break-even demonstrations in the mid-1990s, practitioners produced fundamental advances in plasma processing for electronic materials and developed materials for severe environments. At my next career stop, Jefferson Lab in Newport News, Virginia, the primary mission was fundamental nuclear physics, however, the short-term payoffs were in fields as diverse as radiation protection, medical diagnosis, and powerful lasers (see Chap. 25).

The knowledge gained by conducting basic and applied research in all scientific fields offers payoffs in every aspect of our daily lives. There is no simple formula for predicting the value and impact of new scientific knowledge for future generations. One study [4] estimates the return for each dollar invested in research ranges from 30 to 100%—an impressive return for any investment. For every investment in research—basic or applied, short-term or long-term—we all benefit, both tangibly and intellectually. That is why scientists and members of the research community have to be engaged with the decision-makers within their governments and funding agencies. Decisions made within these deliberations need to be informed, because they will determine the pace of discoveries in coming years.

References

1. Oral history interview with Victor Weisskopf, conducted by Charles Weiner and Gloria Lubkin in 1966, is available through AIP's Niels Bohr Library and Archives; https://www.aip.org/history-programs/niels-bohr-library/oral-histories/4945-1
2. M. Hourihan and D. Parkes, *"Federal R&D Budget Trends: A Short Summary"* AAAS, Washington, DC (2019); https://www.aaas.org/sites/default/files/2019-01/AAAS%20RD%20Primer%202019_2.pdf
3. *"The Future Postponed"*, MIT Report, MIT, Cambridge, MA (2015); https://dc.mit.edu/sites/default/files/Future%20Postponed.pdf, and a follow-up report entitled *"The Future Postponed 2.0"*, MIT, Cambridge, MA (2017); http://www.futurepostponed.org
4. S. Pool and J. Erickson, *"The High Return on Investment for Publicly Funded Research"*, Center for American Progress, Washington, DC (2012); https://www.americanprogress.org/issues/economy/reports/2012/12/10/47481/the-high-return-on-investment-for-publicly-funded-research/

31

Science and Engineering: Hand-in-Hand

The January 2009 inauguration of Barack Obama as the 44th president of the United States rejuvenated a national discussion of the value of scientific research for the common good. Obama began his presidency by affirming "we will restore science to its rightful place" in the nation despite facing the ravages to the national and worldwide economy brought on by the "2008 Great Recession." Early in Obama's term he oversaw and signed a momentous $787 (USD) billion economic stimulus bill [1] that helped the nation emerge from this recession. As this bill proceeded through Congress there were numerous discussions among the scientific community about how science could contribute to the recovery. Early in these discussions, Henry Petroski, an engineering professor and historian at Duke University, published a commentary piece in the *Washington Post* [2] offering a unique perspective on how the nation should proceed with major infrastructure initiatives. Petroski is well known for popularizing engineering with books illustrating a wide range of engineering problems from the design of a pencil to solving the problems of an aging large-scale infrastructure.

Petroski began by quoting a line from President Obama's inaugural speech about harnessing "the sun and the winds and the soil to fuel our cars and run our factories." If we really want to do that, Petroski wrote, "we shouldn't look to science" because "what we need is engineering." Using the example of steam engines—which did see extensive practical use even before scientists sorted out the science of thermodynamics—Petroski asserted that "the truth is that full scientific understanding isn't always necessary for technological advancement." In fact, we need both scientific research and engineering, as Petroski indirectly acknowledged in concluding "Obama should keep his

© Springer Nature Switzerland AG 2020
H. Frederick. Dylla, *Scientific Journeys*,
https://doi.org/10.1007/978-3-030-55800-0_31

promise to 'restore science to its rightful place'—and put engineering on at least an equal footing".

I submitted a response to this commentary, saying that although it's true that engineers' unique indispensability needs better recognition, it's *also* true that when it comes to the batteries, solar cells, and fuel cells Petroski cited, even engineers' most innovative efforts—though crucial and needed immediately—can't ultimately achieve the desired technological advancements. This requires a deeper understanding of nature's mysteries, just as was required for inventing microelectronics, DNA sequencing, medical imaging, and other technologies. Only science can deliver that understanding—which engineers can then apply.

Apparently, the *Post*'s editors recognized that Petroski's article needed a clarifying response. They didn't use my letter, but they printed two [3] that sounded much like it. In the first, Michael Kupper of Rockville wrote, in part, that "semiconductors, the basis of today's computer technology, would have never been developed without scientific interest in materials that engineers at the time considered useless." The letter from Andrew J. Lovinger, a member of the National Academy of Engineering, began, "I was dismayed to read Henry Petroski's outlook commentary. When the rest of the world is rapidly catching up to us in science and technology, the last thing our scientists and engineers should be doing is arguing about who is more important." He, too, gave some examples of new scientific knowledge leading to new technology development—and he closed by declaring that "science and engineering are an integrated enterprise and work best in tandem." Point well taken!

Pretroski acknowledged this key partnership when he expanded his editorial into a full-length book [4], giving numerous examples of important technical problems that were solved by science and engineering working hand-in-hand.

References

1. American Recovery and Reinvestment Act of 2009, U.S. Public Law No.111-5, enacted by the U.S. Congress on Feb. 13, 2009 and signed by President Obama on Feb.17, 2009
2. Henry Pretroski, *"Want to engineer real change? Don't ask a scientist"*, Washington Post, January 25, 2009, http://www.washingtonpost.com/wp-dyn/content/article/2009/01/23/AR2009012302351.html?sub=AR
3. Henry Petroski, *"The essential engineer: why science alone will not solve our global problems"*, Knopf Publishing Group, New York (2010)

4. Letters to the *Washington Post* from Michael Kupper and Andrew Lovinger, February 1, 2009, http://www.washingtonpost.com/wp-dyn/content/article/2009/01/31/AR2009013101537.html

32

The Professor is on the Screen

The first set of large-scale experiments with massively online open courses (MOOCs) are under constant evaluation by the academic community. These experiments have ranged from a plethora of single offerings at individual institutions to several well-publicized consortia, such as those founded by MIT and Harvard [1], Stanford [2], and the European Union [3]. Udacity is another popular third-party platform with an international reach, to nearly 200 countries [4]. This mode of free, or nearly free coursework has also been promoted by a number of civic-minded philanthropists such as Michael Saylor, Laurene Powell Jobs, and Bill and Melinda Gates.

Given the necessity to temporarily close most schools and universities for extended periods for mitigation of the worldwide coronavirus pandemic in 2020, this mode of instruction became a necessary alternative to attending a classroom.

This chapter was adapted from an article first published by the author in *Physics Today* **67** (6), 8 (2014).

© Springer Nature Switzerland AG 2020
H. Frederick. Dylla, *Scientific Journeys*,
https://doi.org/10.1007/978-3-030-55800-0_32

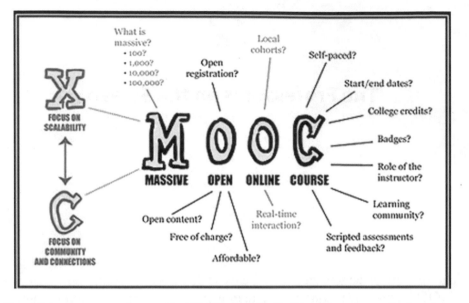

Fig. 32.1 Open questions with MOOCs. *Credit* Mathieu Plourde; mathplourde on Flickr

At first consideration it seems like a wonderful opportunity to be able to take MIT's first-year electrical engineering course or a Stanford course on political science at your leisure at no cost, except for the internet connection and receiving screen.

In 2014, I participated in a forum on the future of engineering education, sponsored by the American Society of Mechanical Engineers and the U.S. National Science Foundation. In a session devoted entirely to the topic of MOOCs, the participants shared some findings from an evaluation of this first wave of MOOC experiments and the potential impact on higher education. Among the data considered was the number of MOOC enrollments compared to the number of completed courses. Registrants for well-known MOOC programs, such as those offered by MIT and Stanford, for example, sometimes exceeded 150,000 per course, whereas the number of students who completed the course requirements was typically less than 5%. On one hand, the low percentage of completion could indicate that success is questionable. However, in terms of reaching individuals, nearly 5% means nearly 7500—via a typical face-to-face course, this could take decades.

Also interesting are the course registrant demographics. For one "innovation and entrepreneurship" course organized and hosted by the Penn State School of Engineering, 80% of the students already had college degrees, more than 70% were international, and—of particular value to the engineering profession where women are seriously under-represented—more than 60% of the registrants were female.

As an experiment in bringing world class higher education content to anyone in the world with interest (and Internet connectivity), these ventures appear successful and should be applauded. Important questions remain pertaining to how this enterprise will mesh with traditional teaching methods on the hosting campus and who will pay for course development and execution, given that most of the registrants are off campus and not paying for services.

This question about the appropriate business model parallels the discourse concerning open access to scholarly publications. There are clear and admirable benefits of distributing the world's knowledge either in courses or in scholarly publications in an open format and available to anyone in the world with a connection to the Internet, but somebody needs to pay to produce and to host this high-value content online.

From the MOOC experiments, there are emerging local benefits for the MOOC producers that may justify the host institution footing the bill for course production. Experiments with so-called "flipped classrooms" are showing the benefits of teachers spending more time with one-on-one exercises or lab projects and leaving the traditional lectures to self-paced online instruction. In such arrangements, the course material prepared for MOOCs becomes an integral or at least an important supplement to the host institution's course material, hence worthy of local investment. In this hybrid format, the on-campus student gets the combined benefit of the professor's (or teaching team's) extensive effort put into developing a high-quality online product with production qualities suitable for a mass audience, in addition to the benefit of personalized face-to-face time with the professor.

The online student does not get the key benefit of the on-campus experience. Creative additions that facilitate learning online with discussion groups and various social media don't yet substitute in my book. I still treasure (some 45 years later) the days I spent in the classrooms, labs, hallways, and professors' offices during my student days at MIT. Perhaps when there are full holographic remote interactions, I will change my mind.

My alma mater was experimenting in online education well before the word MOOC became fashionable in the higher education market. MIT faculty started putting their courses online in 2002; materials from undergraduate and graduate courses alike were made available to the public for free or partly free. Ten years later, almost 2200 of the courses in the hefty MIT course catalog were online.

The proliferation of MOOCs brings up the question of certification. If a bright and adventurous student in a land far, far away takes and "passes" all of the freely available courses for a particular MIT major, could that student be awarded an MIT degree without stepping on campus and without paying tuition (just over $53,000 for the 2019–20 school year)? For now, the answer is no. But many MOOCs do offer certificates of completion.

Yet because MOOCs are still relatively new compared to the century old use of the blackboard lecture, questions about certification/accreditation and other concerns continue to be weighed and evaluated. These digital offerings present both threats and opportunities to institutionalized education; they are also changing the way we receive education and interact with our "classmates" and instructors. I believe that this movement will drive progress and leave some tradition in its wake.

References

1. The edX online course collection: https://www.edx.org
2. The Coursera online course collection: https://www.coursera.org
3. The OpenUpEd online course collection: https://www.openuped.eu
4. The Udacity online course collection: https://www.udacity.com

33

Go Visit a Science Museum

There are two recurrent signs that Spring has arrived in Washington, D.C.—the resplendent cherry blossoms along the Potomac River tidal basin, and throngs of young students and their patient chaperones visiting the Smithsonian's Museum of Natural History. This annual ritual of school field trips, from the first hint of the cherry blossoms until June graduation ceremonies, reminds me of what a superb resource museums and science centers are for all aspects of our culture. It is true that U.S. and many other nation's school systems lack resources for teaching science. We lament over the shortage of properly trained science teachers and insufficient teaching resources that can make science come alive when students see and touch real fossils, charge up their hair with a Van de Graaf generator, or get bedazzled with a planetarium showing our galaxy in a way that can't be seen in most areas because of light pollution. Tapping into the troves of museums and science centers can make a big difference.

In the 1950s when I was a young student, one had to travel to a major city like Washington, D.C., or New York to experience a well-appointed science museum. I attended high school in a rural southern New Jersey school system with few resources to kindle my science interests. Fortunately, I could hop on a Philadelphia-bound bus to spend many afternoons at the Franklin Institute. My teachers would often grant me an extended pass for the day—as long as

© Springer Nature Switzerland AG 2020
H. Frederick. Dylla, *Scientific Journeys*,
https://doi.org/10.1007/978-3-030-55800-0_33

I kept up my grades. I loved this museum and it fed my appetite for science and technology until I left home for college. It is still true that our big cities host the best-resourced and most famous of science museums, but now there are nearly 500 science and technology museums worldwide.

In 2017, the Association of Science-Technology Centers estimates that 45 million people visited 123 science centers and museums in the United States [1]. This volume easily surpasses professional sports game attendance for baseball, basketball, soccer, and ice hockey. Only American football has the museum business beat, but, all told, it's clear that Americans have a healthy appetite for science [2].

In 2015, the U.S. National Science Foundation (NSF) bestowed its Public Service Awards to two of the most outstanding U.S. science centers: the American Museum of Natural History [3] in New York and the Museum of Science [4] in Boston.

The centers are celebrated for their excellence and effectiveness at reaching millions through their rich programs for the public, students, teachers, and science researchers. Millions of U.S. residents, particularly those in the populous mid-Atlantic corridor, pass through their doors each year, but extensive outreach programs enable these establishments to extend their impact to the rest of the country and worldwide.

The American Museum of Natural History, founded in 1869, was recognized as one of the world's most prestigious institutions in its class. In 2015, the year of the NSF award, the museum educated and entertained more than 5 million visitors with 45 distinct exhibits. Traveling exhibits, web exhibits, and even smart phone apps succeeded in doubling its reach. The museum also supports graduate studies in many subjects aligned with its collections and mission, and in 2012 it became the first museum to offer a doctoral program.

The National Center for Technological Literacy (NCTL) [5] at Boston's Museum of Science was recognized by the NSF for bringing engineering and technology demonstrations and exercises to middle schoolers for more than a decade.

Fig. 33.1 The National Center for Technological Literacy at the Museum of Science, Boston, MA. *Credit* Michael Malyszko]

NCTL's dynamic founding director, Ioannis Miaoulis, started its engineering program in 2003 with eight teachers and 200 students and built it up to serve 80,000 teachers and 80 million students a decade later. The Museum of Science has studied and practiced ways of engaging students in middle school, a critical age for developing interest in science, through informal science education.

I encourage everyone, young or old, to visit one of these award-winning institutions—or any one of the other science museums across the world—to enjoy the wonders and beauty of the science and engineering that supports our existence. Better yet, bring along a friend.

References

1. National Science Board selected the American Museum of Natural History for its 2015 Public Service Award; https://www.nsf.gov/news/news_summ.jsp?cntn_id=134879
2. National Science Board selected Museum of Science, Boston's National Center for Technological Literacy for its 2015 Public Service Award; https://www.nsf.gov/news/news_summ.jsp?cntn_id=134950

3. Science Center and Museum Statistics, 2017, Association of Science and Technology Centers; https://www.astc.org/wp-content/uploads/2018/11/ASTC_ScienceCenterStatistics-2017.pdf

4. The spectacle of sports, *The Economist*, June 5, 2014. http://www.economist.com/blogs/graphicdetail/2014/06/daily-chart-2

5. The National Center for Technological Literacy, Museum of Science, Boston, MA; https://legacy.mos.org/nctl/

Part IV
Communicating Science

34

Illuminating Manuscripts

Scholarly journal publishers take pride in adopting web technologies for continuously improving journal production efficiency, content richness, and dissemination methods. For example, the composition language, XML, used to represent a manuscript's digital version, allows all sorts of associated useful data to be "tagged," with the article text and bibliographic data. This optimizes the search function allowing authors, reviewers, and editors to efficiently locate topics or references in the manuscript. As publishers enhance the online user interface for journals and other publications, pull-down menus of information related to the manuscript are being added to aid the reader. Scholarly publishers regard these online features as significant improvements in utility over hard copy—or are we really being a little smug? It turns out a precedent for such innovations existed well before the arrival of Gutenberg's press.

One of civilization's first needs for printed information was a compendium of laws to establish basic rules of conduct. In the fourth century, the Roman emperor Justinian brought into existence the first opus of Roman law, the Corpus Juris Civilis (Body of Civil Laws). Unfortunately, the manuscript has been lost since 603. However, some nameless soul produced a summary of the document, called the Digest—which may be the first example of an abstracting service at work. Unfortunately for the inhabitants of Europe's Middle Ages, only two copies of the Digest were known to have survived, and their whereabouts were unknown for four centuries.

An Italian liberal arts teacher, Irnerius, found a copy of the Digest while working in the Royal Law School in Ravenna, Italy, in 1076. The Digest was a very complex document about an empire that had little bearing on

© Springer Nature Switzerland AG 2020
H. Frederick. Dylla, *Scientific Journeys*,
https://doi.org/10.1007/978-3-030-55800-0_34

Europe at the end of the first millennium. However, Irnerius' gift to Western civilization was his subsequent editorial work. He meticulously "tagged" the Digest by adding a series of explanatory notes in the margins and between the lines of the manuscript. Tagging was already in use by monks in the monasteries of Europe who used the technique to enrich hand-written biblical texts. These tags are easily seen in the beautiful page that is illustrated here from the Epistle of St. Paul produced in the mid-twelfth century.

Fig. 34.1 Page from the epistle of St. Paul, showing tags. *Credit* Bodleian Libraries

Irnerius's version of the Digest became the basis of a teaching system he founded in the nearby St. Stephen monastery. As expounded by James Burke [1], my favorite British scholar of Italian literature and history of science, Irnerius urged his teachers to produce the following aids for their scholars: "summulae" (notes summing up whole areas of law), "continuations" (summaries of different groups of laws), and "distinctions" (variations on the hypothetical cases described). If you substitute the subject of "physics" for "law" in these teaching aids, you will find something similar in the website, *Physics* [2], launched by the American Physical Society in 2008 to explain and summarize the best manuscripts appearing in the *Physical Review* suite of physics journals.

We think we are very smart, but it pays student and publisher alike to check the literature!

References

1 James Burke, *The Day the Universe Changed*, Little, Brown and Company, Boston, 1985, pp. 33–36

2. *Physics*, see https://physics.aps.org

35

Science Must Navigate the Gale of Creative Destruction

Economist Joseph Schumpeter was the first to observe that disruptive but necessary turbulence often drives economic growth [1]. But thanks to the Internet, something like Schumpeter's "perennial gale of creative destruction" is also churning up the way we process and communicate the results of science.

Fig. 35.1 Economist Joseph Schumpeter. *Credit* Volkswirtschaftliches Institut, Germany]

Long before capitalism had fully evolved, science established itself as a collaborative knowledge-generation enterprise that relied, among other practices, on peer-reviewed publications as the primary form of communication.

© Springer Nature Switzerland AG 2020
H. Frederick. Dylla, *Scientific Journeys*,
https://doi.org/10.1007/978-3-030-55800-0_35

What researchers reported in scientific papers and books could be tested, criticized, and refined. Once established, new knowledge was used for developing even newer knowledge.

Sometimes, as with Roentgen and X-rays, or Fleming and penicillin, the new knowledge could be applied in practical ways. In the twentieth century, with overlaps growing between pure and applied research and between research and engineering, science came to drive the advance of technology.

Obviously, that meant science also came to contribute mightily to Schumpeter's gale, and thereby to prosperity and human well-being. Creative destruction occurred when electricity put people out of work in the candle and oil-lamp industries, when jet airliners ended transoceanic ship travel, and when computers killed typewriters.

Science still advances through communication and collaboration, but now the Internet offers new ways for it to do so. Science needs not just to accept and actively guide the inevitable changes. That's why it's important to declare that trying to shelter the scientific publishing industry from the Internet's relentless opportunities is counterproductive resistance to Schumpeter's gale.

Worldwide, over many decades, scientific publishing has grown to multibillion-dollar scale [2].

The business spans a wide variety of organizations including nonprofits, scholarly society-based publishers, and large commercial firms. Under copyright protection, scholarly publishers each year transform more than two million scientists' manuscripts into peer-reviewed articles, professionally edited, produced, and archived. Publishers make these articles available primarily via online journal subscriptions to libraries, and usually with a liberal interpretation of "fair use," allowing authors to post articles on their websites and distribute copies for noncommercial uses.

But worldwide in recent years, people in and out of science have been asking why scientific papers must be sequestered behind journals' subscription paywalls. Why not just have scientists post their manuscripts online, publicize the links, let web search engines handle information retrieval, and be done with it? It's a great question reflecting science's natural, and crucial, spirit of openness. Consider why scientists at Geneva's CERN physics laboratory devised the World Wide Web in the first place. They wanted better, simpler ways to use the Internet for collaborating in the advance of particle physics.

The question carries special urgency among nonscientists who hope, sometimes desperately, to search medical knowledge for ways of combating loved ones' illnesses. It also animates researchers who are impatient to see their own and others' papers made quickly and easily available free of charge online.

After all, the most vocal proponents assert, taxpayers foot the bill for much of what scientists do, don't they? So why should taxpayers pay again to see research results? The simple answer to this question is yes, a public agency may have paid for the research, but that agency has not paid for the added value that a scholarly publisher invests in the final peer-reviewed article.

This issue has a name: *open access*. Ironically, scientists themselves treasure the principle of free-flowing information. We can't have science without it. I see this outlook in every scientist I know in scholarly publishing. We want to see the scientific literature opened up online for access beyond what's available to users of university and laboratory libraries. But we also want to see science continue evolving sensibly and effectively as a knowledge-generation enterprise. After all, free access can't be costless. It takes a few thousand dollars apiece to transform manuscripts into what science continues to need: formally vetted (by peer and editorial review), published articles, archived for efficient information retrieval. Secondary school students are taught to be careful in selecting information sources on the raucous, ungoverned Internet. Doesn't the same dictum apply even more for the scientific literature?

None of this means though, that in science or in scientific publishing Schumpeter's gale should be resisted. It does mean working hard to make sure that modern science—a vessel for knowledge that took over 300 years to create—doesn't founder during the continuous transition to better ways of promulgating and sharing information.

References

1. Joseph A Schumpeter, *"Business Cycles: A Theoretical, Historical, and Statistical Analysis of the Capitalist Process"*, McGraw Hill, New York (1939)
2. Robert Campbell, Ed Pentz, and Ian Borthwick, eds, *"Academic and Professional Publishing"*, Chandos Publishing, Oxford, UK (2012)

36

Roundtables Help

Scholarly publications, such as journals, are the primary form of communicating scientific research and have served this function well for the past 360 years. Since the first journals were founded by learned societies in Britain and France in the late seventeenth Century, subscription income has sustained the scholarly publication system and the activities of many learned societies who publish journals. With the advent of the Web, it became easy for authors to post their work onto to some Web platform as soon as the written version was complete. That content, however, is not likely to be accepted into the body of scholarly knowledge without essential contributions usually provided by a publisher versed in the author's field of study. The most important contribution provided by the publisher is the organization of the peer review for the article, helping ensure the quality of the research and its presentation to readers. The publisher also provides wide dissemination in a variety of formats, as well as archiving journal content. These latter two functions are shared with scholarly publishers' primary customer—the research library. The cost of these functional roles of publishing (peer reviewing, formatting, posting, distributing, archiving, etc.) is typically between $1000 and $4000 per article and, in certain cases, substantially more. Such costs are largely recovered through institutional library subscriptions and license fees in the case of more than 80% of the global scholarly publication business, despite the growth of an author or sponsor-paid "open access" business model that first appeared in the early 2000s.

Since the advent of online journal access in the mid 1990s, access by individual readers has exploded. In the print era, printed copies of journals were

© Springer Nature Switzerland AG 2020
H. Frederick. Dylla, *Scientific Journeys*,
https://doi.org/10.1007/978-3-030-55800-0_36

delivered primarily to one or more libraries at a research institution or university and there were very few individual subscribers. When journals shifted to online access, anyone associated with the research institution could be delivered a copy to their desktop computer. Scholarly publishers and librarians have worked together to provide greater access to an increasing number of journals and their old print-files now digitized, allowing scholars around the world to access, download, read, and interact with more peer-reviewed scholarly literature.

Some believe strongly that scholarly publications should be available at no charge to everyone with a web connection, asserting that this would provide wide benefit to science and its applications, particularly in the case of less advantaged institutions and small businesses around the world. Furthermore, the argument goes, the results of research funded by the public (taxpayers) should be freely available to the public. Indeed, this would be ideal, but such a gift to public readers comes with a price because there are real costs to be accounted for—as the stressed music recording and newspaper businesses found when those industries transitioned to online business models.

I have thought a lot about the cost of providing high quality scholarly communications over my decade-long career as the CEO of the American Institute of Physics (AIP) with its significant physics journal portfolio. My thoughts on this subject led to series of arcane talks and papers at scientific conferences as well as a number of general media contributions [1–3]. I found the latter important to do because, despite the fact most scientists spend an important part of their career publishing their results in scholarly journals for their scientific peers, scientists are generally very poor at communicating the importance and value of science to the general public, who support much of this research through their tax payments.

During my research career, my institutional librarian was often a key partner in my research efforts. I was dismayed as I began a stint as a scientific publisher to find myself on the other side of the debate with some research librarians about journal access. To help bridge the divide in this debate regarding journal access among the stakeholders (but particularly among publishers and librarians), I was asked by a member of the U.S. Congress to help organize and convene the "Scholarly Publishing Roundtable" in 2009–2010. This group's charge was to provide recommendations to the federal government on how to accomplish public access to the results of federally funded research–specifically journal articles and associated data. The Roundtable members were drawn from the library, university administration, research, and publishing communities. After six months of sequestered deliberation, the group released its report to the U.S. Congress in early 2010

[4]. The core recommendation recognizes the responsibility of the federal government to develop such policy for publicly funded research. However, the Roundtable urged the government to work in collaboration with all stakeholders in order to ensure the sustainability of the scholarly journal enterprise. In other words, the Roundtable urged all stakeholders to continue to act as the Roundtable members did–in collaboration and with recognition of differing views of the diverse stakeholders.

Roundtable members confirmed the key tasks of scholarly publishing: the necessity of maintaining a viable and independent peer review system, increasing access, adaptable business models, interoperability of online platforms, and the need to pay careful attention to archiving and preservation. The group agreed scholarly publications are essential to scholarship and any changes to the system of scholarly publishing would have to be carefully designed, implemented and monitored (i.e., improve the system but don't break it). The desired interoperability among all publishing and archival platforms cannot be fully realized through a single government, institution, or publisher platform. It requires a cooperative system.

One of the most important outcomes of the Roundtable transcends any of the individual recommendations in the report. Members of the Roundtable uniformly felt that a useful process for dealing with a contentious issue among stakeholders was demonstrated. And, while not arriving at an absolute consensus, the overwhelming majority of the Roundtable members (12 out of 14) endorsed the full report and its recommendations. Members came to the first meeting with a diverse set of views, but by the time the report was completed a set of recommendations had been fashioned that required compromise by all parties. Members also left the Roundtable with a conviction that many of the recommendations and the process employed will help deliver a sustainable path for the evolution of scholarly publications. A decade later, Roundtable members were pleased to see that their efforts to steer a path through a contentious issue had paid off.

Just a year after the Roundtable report was delivered to the U.S. Congress, the key recommendations ended up in legislation that was passed by Congress and signed into law by President Obama in January of 2011 [5]. Two years later, in February, 2013, the President's Science Advisor, John Holdren, issued a directive [6] to all of the major federal government science agencies instructing them to implement a plan for public access to publications and data resulting from federally funded research. Coincident with this presidential directive, a group of scientific publishers (including AIP) founded a non-profit organization, CHORUS [7, 8], to assist U.S funding agencies in the implementation of this plan. By the end of 2018, public access to more than 90% of publications describing U.S. agency research was enabled. The access debate is far from over given the fast pace of web-based business models, but the Roundtable showed the value of careful consensus-building when all parties are represented and given an environment that fosters and values give-and-take.

References

1. America COMPETES Reauthorization Act of 2010, Public Law 111-358, January 4, 2011, https://www.congress.gov/111/plaws/publ358/PLAW-111publ358.pdf
2. CHORUS, the Clearing House for the Research of the United States; https://chorus access.org
3. H. Frederick Dylla, "*A defense for the high cost of many scientific journals*", letter published in *The Wall Street Journal*, April 18,2016: https://www.wsj.com/articles/a-defense-for-the-high-cost-of-many-scientific-journals-1461000697
4. H. Frederick Dylla, "*Real threat to research access is stagnation in public funding*", letter published in *The Chronicle of Higher Education*, May 26, 2016: https://www.chronicle.com/blogs/letters/real-threat-to-research-access-is-stagnation-of-public-funding/
5. H. Frederick Dylla, "*Academic Publishing*", letter published in *The Economist*, August 04,2012, https://www.economist.com/letters/2012/08/04/on-fracking-robots-gun-laws-the-music-industry-education-milton-friedman-plurals-academic-pub lishing-kim-jong-uns-wife
6. H. Frederick Dylla and Jeffery Salmon, "*Collaborating for public access to scholarly publications: A case study of the partnership between the US Department of Energy and CHORUS*", *Learned Publishing*, April, 2020; https://doi/10.1002/leap.1298
7. Report from the Scholarly Publishing Roundtable, Association of American Universities, Washington, DC, January 2010, https://www.aau.edu/sites/default/files/AAU%20Files/Key%20Issues/Intellectual%20Property/Scholarly%20Publ ishing%20Roundtable%20Report%20and%20Recommendations%20-%201-12-10.pdf
8. U.S. Whitehouse Office of Science and Technology Policy memorandum on "*Increasing access to the results of federally funded scientific research*", February 22, 2013, https://obamawhitehouse.archives.gov/sites/default/files/microsites/ostp/ostp_public_access_memo_2013.pdf

37

Imagine a World Without Editors

Over my decade-long career sojourn as a scientific publisher, I never imagined that I could have more appreciation than I have always had for the editing process. After all, the distillation of overwhelming quantities of accessible information is vitally important to the publishing industry. But then I heard acclaimed author Malcolm Gladwell give an address to a group of publishers meeting in New York City [1]. Gladwell's talk centered on the value editors bring to scholarly publishing; they separate the valuable from the less valuable, with the potential to make what is good text great. The editorial process preserves the value and integrity of published works.

Fig. 37.1 Malcolm Gladwell. *Credit* Kris Krüg, flickr, CC-BY-2.0

© Springer Nature Switzerland AG 2020
H. Frederick. Dylla, *Scientific Journeys*,
https://doi.org/10.1007/978-3-030-55800-0_37

Gladwell, who has four *New York Times* best sellers to his credit, made the intriguing assertion that the late Steve Jobs—the high-tech guru and Apple products creator—was not distinguished as the creator of new technology, but as someone who improved technology created by others, making it accessible and appealing to potential consumers.

In the publishing world, whether it is a short story, a book-length collection of works, or a journal article, a good editor saves valuable time and makes a literary work more user-friendly by directing our attention to the core information and salient points. With the advent of the Internet, vast information archives are now freely available to everyone. But to borrow a phrase from the early days of the Internet, the information superhighway is seriously overcrowded to the point of data gridlock. The most daunting problem for the casual information seeker and the most serious scholar is not the lack of ready access to vast quantities of information, but rather the effective editing of the irrelevant from what is most desired and helpful.

Gladwell creatively illustrated the value of the editing process by noting examples where modern societies have not sufficiently applied the benefits of editing to other disciplines. He pointed out the quandary that exists in modern medicine, wherein both detection and treatment methods for certain chronic diseases have increased significantly—there's a lot of new information out there, but the success rate for the prevention of premature deaths has not improved over the last 50 years.

Expert editing is the catalyst that will push the publishing industry forward. "Don't give me more," said Gladwell, "Give me less and make it good, and you'll be in business forever." These days, just about anybody can write and even self-publish a book, but a talented author matched with an experienced editor can write a book worth reading over and over. Ernest Hemingway and Marjorie Kinnan Rawlings were great writers, but their greatest works were published before 1947, which was the year their editor, Max Perkins, died.

So what is the connection to my sojourn in the business of scholarly publishing? Scholarly journal editors are the first to look at submitted articles, which are then passed onto peer reviewers who perform essential quality checks. Editors are often called upon to adjudicate the views of several reviewers before a peer-reviewed article can be published or rejected.

There is clearly too much data residing on web-based platforms for a working scientist to effectively review. Therefore, it is incumbent upon scholarly publishers, together with information and data scientists, to continually develop and improve new search and discovery tools that navigate on search terms tagged within the text to allow the reader to find the essential information they want. The information superhighway is moving into its fifth decade of existence. The task before the highway builders is better road maps.

Reference

1. Annual meeting of the Association of American Publishers, March 2012, New York City

38

Please Read My Paper

There is no shortage of reading material in the research community. Scholars publish in English nearly three million journal articles a year in over 33,000 scholarly journals [1]. This is the result of a well-established trend. The number of journal titles has grown roughly 3% a year since the first scholarly journals appeared over three and a half centuries ago. In 1665, The British Royal Society began to publish *Philosophical Transactions* [2], the same year that the French Académie Royale commenced publication of *Journal des Sçavans*. As pointed out in analysis by Mabe [3], this growth in journals simply scales with the number of researchers. There is no way for an individual to deal with the volume of content, even a subset restricted to a field of the literature that matches the researcher's interest [4]. Regardless, access to the scholarly literature has become a high-profile issue in the academic community because much of the literature is supported by subscription access. The more acute problem because of the immense and growing content of the scholarly literature is discovery of the most important articles—and its associated data—by the inquisitive researcher. How can one identify the absolutely essential research articles pertinent to any particular research endeavor?

To solve this problem, we will not rely on an army of research assistants, because even the low-cost, extended workday of graduate students can't make a dent in this problem. Clearly, the dominant reader of the future will be a machine rather than a person. Adapting the scholarly literature to be efficiently and accurately machine-readable and developing machine-reading tools with user-friendly interfaces are frontier development projects in the publishing and information technology communities. This enterprise has a

© Springer Nature Switzerland AG 2020
H. Frederick. Dylla, *Scientific Journeys*,
https://doi.org/10.1007/978-3-030-55800-0_38

catchy name—text and data mining, or TDM for short—and there is considerable discussion of its prospects and potential benefits in the publishing community and among its customers and policy makers.

In its simplest form, the search and indexing routines used by commercial search engines such as Google and Bing perform text mining adopted to the full corpus of literature on the web to allow key topics to be discovered and exposed by these search engines. Most scholarly publishers sign agreements with these firms to allow their content to be "crawled" by robotic readers to tag the content for identification using key words and terms. Taken to the next level, more sophisticated TDM uses more sophisticated analytic tools (such as natural language processing and machine learning) to recognize relationships in unstructured text and key identifiers, such as names, chemical structures, and experimental methods. TDM is an arena ripe for research, development, and testing of techniques. But there are also cautions that need to be heeded as machine reading evolves as a widespread and necessary tool. All public and private databases of digital content have to be protected from the ubiquitous online threats.

The potential value of TDM tools and techniques is greatly enhanced if the widest possible collection of all content, from related and seemingly unrelated subjects, are made available for the mining exercise. It opens up the possibility of serendipitous discoveries when connections or relationships are examined beyond narrower searches. Within the realm of physics, there is the example of 2013 Physics Nobel Prize for the discovery of the Higgs boson (see Chap. 23). The fundamental theoretical work by Peter Higgs and his collaborators in high energy physics was based on examination of quantum phenomena in superconductors previously done by Anderson [5], a prior (1977) Nobel prize winner in theoretical solid-state physics.

We can look forward to more purposeful and serendipitous discoveries from TDM as the technique becomes a common tool of research and scholarship. One of the most active applications is in the fields of biomedical and pharmaceutical research—even before the impetus given by the worldwide coronavirus pandemic in 2020. Important topics such as drug discovery and patient reactions can be tracked (with non-trivial effort and necessary care for patient privacy) across the diverse array of medical records, drug and patient trials, and medical journal literature. Information that can be quickly targeted to a specific query, easily accessed, traced to reputable sources, and widely disseminated can save lives.

References

1. Mark Ware and Michael Mabe, "*The STM Report: an overview of scientific and scholarly publishing, 2018*", STM Publishing Association, London, 2018; https://www.stm-assoc.org/2018_10_04_STM_Report_2018.pdf
2. See Chapter 41 for an account of the birth of the first scientific journal
3. Michael Mabe, *Serials* **16**(2), 191–7, 2003
4. A.G. Fraser and F.D. Dunstan, "*On the impossibility of being an expert*", *BMJ* 2010;341:c6815
5. P.W. Anderson, *Physical Review* **150**, 439, 1963

39

Getting It Right

Are scholarly publications any more reliable or trustworthy than the general press? Does the scholarly publishing process build in sufficient checks and balances or is it remarkably flawed? The occasional reports of fraudulent or retracted articles from the body of scholarly literature might suggest a crisis in scholarly publishing. The organization Retraction Watch [1], founded by Ivan Oransky and Adam Marcus, tracks scholarly journal articles that have been retracted for any reason. Some are retracted where no misdeed has been committed, such as an author finding mistakes in an original submission. Some retractions are forced because the underlying science or its reporting was indeed fraudulent.

Airing out this dirty laundry keeps the scholarly publishing industry in check, but unfortunately also occasionally attracts the attention of the popular press and its predictable fallout of sensationalism. A cover story in *The Economist*, "*Unreliable Research: Trouble at the Lab*" [2], in 2012 is one highly cited example. Stories on so-called "predatory publishers" and the rising number of retracted articles support the not-so-veiled accusation that the scholarly publishing business has serious quality-control problems, if not unbridled issues with integrity.

Retraction Watch is a valuable auditing service for the scholarly journal business. Without question, a fully retracted article is a sign of the most egregious error that can occur in this form of communication. Reputable journals post and archive an article's official version of record and append any errata or notice of retraction that may subsequently occur. *Retraction Watch* adds another layer of transparency to this error correction. In addition, since *Retraction Watch* covers all fields of scholarly publication, it provides

© Springer Nature Switzerland AG 2020
H. Frederick. Dylla, *Scientific Journeys*,
https://doi.org/10.1007/978-3-030-55800-0_39

some measure of industry-wide statistics. I observe that the numbers are telling in that they are very small in comparison to publication totals. For example, *Retraction Watch* posted approximately 500 retracted articles in the year after the publication of *The Economist* article. Compare this number to the nearly 2 million articles that were published in the prior year (2012) in more than 28,000 scholarly publications. That puts fully retracted articles at about 0.02% of the annual publication volume. Also of note is the fact that the large majority of the retractions listed in *Retraction Watch* database are in biomedical or clinical fields. These research areas clearly have more difficult problems in establishing reproducible starting conditions (cell lines, animal cohorts, well-characterized reagents, etc.) than other areas within the physical sciences. Practitioners in medical fields are aware of these problems and are taking steps to improve the reproducibility of experimentation by redoubling certification and testing procedures.

There have been a number of sting operations to test the publishing process and expose unethical publishers who publish bogus articles that have attributions of scholarly work. John Bohannon, an investigative reporter for *Science* magazine, wrote about a sting operation that he devised to uncover a cohort of largely "pay-to-play" publishers willing to publish almost anything vaguely scientific as long as the author paid a publishing fee [3]. Bohannon submitted a paper on completely fabricated research to more than 300 journals around the world. 255 responded and an astounding 157 accepted the manuscript for publication. Most of these publishers were new, hailing from developing countries, but a few were established and respected in the industry. Bohannon tracked his faux article from submission, to author payment, to acceptance by the publisher. He also traced facade addresses in Western locations and circuitous routes that the payments were funneled to hide the identity of certain publishers. By publicizing the existence of these shadowy enterprises and exposing their methods, he did the industry, authors, and readers a great service. But the success of this venture may have biased his vision as an investigative reporter, as he expressed his own personal distrust of the entire scholarly publishing business.

I stress it is important to take a step back and see these deficiencies *in context*. *Quantifying* retractions, such as offered by *Retraction Watch*, gives good definition to the problem. By *categorizing* those publishers exposed by Bohannon with his sting operation, we see that the great majority are new to scholarly publishing and are from developing countries where the tradition of industry integrity is not yet ingrained.

The scholarly publishing system is not perfect, but errata and retraction statistics belie a miniscule error rate compared to any other communications media. More importantly, the system is self-correcting. It may take time, but bad science or fraudulent science is eventually smoked out—from the fraud of the Piltdown Man [4] in the early twentieth Century to the Hendrik Schön affair [5] at Bell Labs in the first decade of the twenty-first century. As a whole, scholarly publishers and the academic community practice due diligence to maintain the integrity of published works.

Most medium to large scale scholarly publishers conduct annual ethics audits to assure that their policies and procedures are well followed and effective in ferreting out misconduct and substandard manuscripts. The Committee on Publication Ethics (COPE) [6], is a forum for editors and publishers of peer-reviewed journals to discuss and advise on publication ethics. More than 12,000 journal editors and publishers from around the world abide by the principles of operation developed by COPE. By belonging to COPE, editors and publishers recognize that the system is imperfect, but they nevertheless have faith in it and are committed to continually improving it. Science depends on this commitment.

Every discipline has a bad actor or two. The scientific community, as a whole, has devised a methodology to detect and address faulty science. Exercise of this method may cause occasional publicity problems but it keeps the scientific record self-correcting.

References

1. Retraction Watch; http://retractionwatch.com
2. "*Trouble at the lab*", *The Economist*, October 18, 2013; https://www.economist.com/briefing/2013/10/18/trouble-at-the-lab
3. John Bohannon, "*Who's Afraid of Peer Review*", *Science* **342**, 6154, 60 (2013); https://science.sciencemag.org/content/342/6154/60.full
4. John E. Walsh, "*Unraveling Piltdown: The Science Fraud of the Century and Its Solution*", Random House, New York (1996)
5. "*Physicist found guilty of misconduct*", *Nature*, September 26, 2002; https://doi.org/10.1038/news020923-9
6. COPE: Committee on Publication Ethics; https://publicationethics.org

40

A Vision for Open Scholarship

Because of their role in inventing computing technologies, scientists and engineers often feel they play the lead role with innovative uses of these technologies. We should dispense with this limiting view, remove our blinders, and take careful note of our friends in the humanities who are leading an Internet-age initiative that merits science's attention and participation.

In an article in the *New York Review of Books* [1], Robert Darnton—a Rhodes scholar who became a distinguished historian and then Harvard University's library director—presented "The National Digital Public Library is Launched!" That exclamation point in the headline reflects Darnton's optimism about transforming the information environment via the Digital Public Library of America (DPLA) [2].

Though the essay never specifically mentions scientific publications, it calls to mind the brightest Internet-age possibilities for science communication, which have been enhanced considerably by the move from the static printed page to the flashing and interlinked images on a screen. No one yet knows exactly how the evolving communication prospects for science intersect with or complement the DPLA's promise, but the question invites thought.

The DPLA "*harkens back to the eighteenth century*" and the Enlightenment's faith in the free flow of ideas, Darnton wrote. He invoked the scientist and statesman Benjamin Franklin and the science-minded Thomas Jefferson—who, even while serving as U.S. Vice President and then President, presided over the early republic's leading institution for communicating scientific advances, the American Philosophical Society.

The DPLA's founders have resolved to build and maintain "*an open, distributed network of comprehensive online resources…in order to educate,*

© Springer Nature Switzerland AG 2020
H. Frederick. Dylla, *Scientific Journeys*,
https://doi.org/10.1007/978-3-030-55800-0_40

inform, and empower everyone." A closely related resolve permeates science: the intention to increase public access to the scientific literature online. In effect, Darnton alluded to public access in his opening line, which called the DPLA "*a project to make the holdings of America's research libraries, archives, and museums available to all Americans—and eventually to everyone in the world—online and free of charge.*"

But as has been discussed frequently in the two-decade movement for open access publishing, "free of charge" cannot mean "free of cost." That's why scientists' open-access vision directly reflects the tension in what Darnton called "*two currents that have shaped American civilization: utopianism and pragmatism.*" For centuries science's communication system has evolved with publishers adding substantial value to researchers' manuscripts and to the research enterprise itself. Online journal distribution has significantly lowered distribution cost to a much wider audience, but the cost of producing the so-called first copy has actually increased because of the rich and useful complexity of electronic formats.

Like the DPLA, which Darnton attributed to "*a grand coalition of foundations and research libraries,*" scientific societies are working with scholarly publishers, libraries, research funders, and others to find pragmatic ways to realize this vision of the widest possible access. In my view Darnton is right that we now "*have the technological and economic resources to make all the collections of all our libraries accessible to all our fellow citizens.*" It's a grand vision that scientists share, but it's important to remember that science's part must follow a careful pragmatic path forward in order to not jeopardize the essential independent review that the scholarly journal enterprise brings to science.

References

1. Robert Darnton, "*The National Digital Public Library is launched!*", New York Review of Books, April 25, 2013, New York; https://www.nybooks.com/articles/2013/04/25/national-digital-public-library-launched/
2. Digital Public Library of America; https://dp.la

41

The Scientific Journal Marks 350

On March 6, 2015, the scientific community celebrated the 350th anniversary of the founding of the scientific journal. On that date in London, Henry Oldenburg, the first secretary of the Royal Society, an organization founded by Royal Charter less than five years earlier, began publishing the *Philosophical Transactions of the Royal Society* [1]. This publication looked remarkably like the present day scholarly journal—the format still recognized and used by the scientific community as the primary means of communicating scientific results.

The first issue of *Transactions* was a bound collection of articles submitted by Royal Society members. The authors were largely physicians or "natural philosophers." This latter title predated the emergence of the common use of the modern label of "scientist" by almost two centuries. The articles in *Transactions* led off with an explanatory summary still called an abstract. This was followed by the main body of text, explaining the nature of the study and conclusions, often with included figures and annotated references. More remarkable than the constancy of the article format was the early establishment of the key journal article benefits, both for the authors and the reading community. These included: scientific discovery registration date, certification (peer review of the content before publication was granted), distribution of the printed volume as a means of disseminating the results well beyond the group of contributing authors and archiving of the published content. These are all still considered the essential values of the scholarly journal today.

© Springer Nature Switzerland AG 2020
H. Frederick. Dylla, *Scientific Journeys*,
https://doi.org/10.1007/978-3-030-55800-0_41

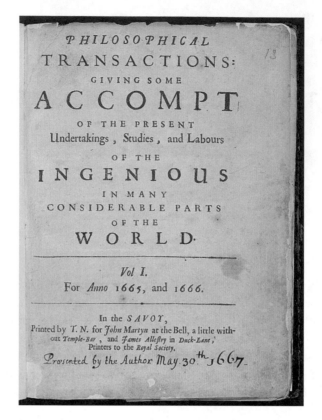

Fig. 41.1 Frontispiece from volume 1 of the *Philosophical Transactions of the Royal Society*, London, 1665–66. *Credit* Henry Oldenburg for the Royal Society, Wikipedia Commons, CC-BY-4.0

The first journal even had a business model that is largely intact to this day. Oldenburg was given a contract by the Royal Society's president, Robert Boyle, to publish and distribute the journal at his expense with the incentive that he could keep any profits from the venture. Oldenburg was required to give free copies of the *Transactions* to all members of the Royal Society, but he was free to sell subscriptions to non-members, thus beginning the journal subscription model. Unfortunately, during the entire 12 years that he was publisher he never recovered his costs—a result typical of many entrepreneurial ventures. In fact, the journal didn't start to be profitable until after 1948 when a significant number of institutional subscribers were attracted from international venues beyond the U.K. The long gestation of the subscription model may contribute to its resilience, since this model is still the dominant source of income for the sustainability of many scholarly journals. The *Philosophical Transactions of the Royal Society* still exists today as the oldest continuously published journal.

The original cataloging system of *Transactions*, with articles bundled into issues and then into volumes, also remains the norm for many journals. However, different forms of journal articles have also appeared over this 350-year evolution, for example, briefer formats such as letters, first appearing in the second volume of *Transactions*, and extended formats such as review articles that sometimes make up entire journals. Virtually every field and subfield have an associated journal, as the number of titles has grown to its present volume of more than 33,000 titles in the English language, published by more than 4,000 scholarly publishers [2]. Throughout this growth of journals, the essential peer review process largely remains as it appeared in the first journal. There have been a number of experiments in the online journal era with post-publication peer review after articles are posted online as submitted, but these experiments don't appear to have gained much traction for replacing the traditional system.

Even though the essential format of the modern journal remains similar to the original Oldenburg model, the migration of the format from print to the online environment has allowed the format to expand in utility. In particular, the innovation of reference tagging and article linking with attached "digital object identifiers," pioneered by the Crossref [3] organization, is a significant boon to identifying and tracking articles. Crossref's present registry of online content exceeded 100 million digital objects in 2018, and its content grows by nearly two million new articles each year.

By moving away from the cost of a printed page, the expanded online format allows all sorts of added features to be associated with the basic article, from underlying tagging that again enhances discoverability of specific content in the article, to associated data and supplementary material to the primary journal article. These enhancements have helped, but not solved, the problem of how scientists keep up with the explosive amount of published content each year. It remains to be seen, 25 years after the first scientific journals went online, whether the basic structure of the journal article will be deconstructed into a completely new format. Although that hasn't happened yet, publishers and entrepreneurs in the trade continue to experiment with journal structures and related content to satisfy the demands of their authors and readers. I venture to say that the next few decades will bring significant change to this time-honored tradition of disseminating scholarly research.

References

1. For a lively account of Henry Oldenburg and the first journal see: A. Singleton, *Learned Publishing* 2014, **27**: 2–4 https://doi.org/10.1087/20140101.
2. Mark Ware and Michael Mabe, "*The STM Report: an overview of scientific and scholarly publishing, 2018*", STM Publishing Association, London, 2018; https://www.stm-assoc.org/2018_10_04_STM_Report_2018.pdf
3. Crossref; https://www.crossref.org

Part V
Art and Science

42

Time Exposed for Science

Young minds are by nature inquisitive, so exposure to the mysteries of science often sparks the desire to learn. Certainly, the dawn of the "space age" had a significant influence on my generation, then in grammar school—we could almost glimpse our future with the first satellite launch. Yet, equally influential to sustaining interest in science at the time was the U.S. government's increasing support for science, both in terms of funding research and improving science and engineering education.

What followed were decades of science proliferation, both public and private. For example, America's blue-chip companies displayed their R&D prowess at the 1964 World's Fair, and those same companies sponsored visits to corporate laboratories for promising high school students. The National Science Foundation sponsored summer-long science camps at more than 1000 college campuses, and I was fortunate enough to participate. There were ample activities, such as those that supplemented my formal pre-college education. With the increasing globalization of industry in the 1970s and the end of the Cold War in the late 1980s, however, the nation's ability to provide what had seemed a boundless and energetic support of science began to wane for all but biomedicine. Since the golden age of the post-Sputnik era, there have been many, not wholly successful, attempts to revive this apolitical spirit.

But regardless of the time in which we grew up, nature itself exposes us to mysteries that beg to be uncovered—we only have to look at the rocks beneath our feet. True, bedrock that may be exposed at the side of a road or pebbles at a river's edge may seem lackluster at first glance, but taking the time to explore more closely, a whole world of color and structure enfolds.

© Springer Nature Switzerland AG 2020
H. Frederick. Dylla, *Scientific Journeys*,
https://doi.org/10.1007/978-3-030-55800-0_42

For me, it was what I picked up off the ground that really ignited my interest in science: the first time I cracked open a rock and found sparkling crystals inside; my delusions of striking it rich when I saw my first flakes of "fool's gold" in a vein of quartz spiraling through a boulder; picking up a dirty piece of shale in a shallow creek bed and finding a fossilized fern entombed for millions of years in an ancient forest floor staring back at me.

Observing rocks and minerals compelled me to learn more: where could I find some really big crystals, and why did they grow in such beautiful and regular patterns? I developed interests in chemistry and biology as I tried to identify weird specimens that didn't match simple descriptions in a book, and an interest in physics as I picked up minerals that were radioactive and fluoresced under ultraviolet light. Even before I went to high school, I knew that I had to become a physicist and get to the heart of the structure of matter. And I still value what's become a lifelong interest in minerals, observing their microstructure under a microscope and their macrostructure through an airplane window.

Fig. 42.1 Photographer Norman Barker, with two images from his collection, *Ancient Microworlds*. *Credit* American Institute of Physics

Visitors who came to the 2014 opening of the art exhibit *Time Exposed* at the American Center for Physics [1] were treated to a selection of photographs from the mineral world that enlivened appreciation for natural wonder. Norman Barker, who has an interesting, juxtaposed title as professor of pathology and art at Johns Hopkins University, has devoted more than a decade to composing photographs of fossils visualized under low magnification (10×) with illumination that accentuates the bright minerals that replace living matter as it fossilizes [2].

I find the exhibit a perfect matchup of art and science. The photographs play with light and color to capture the beauty of natural structures, immediately stimulating the visual cortex. With the captions provided by Barker and his collaborator at Johns Hopkins, Dr. Giraud Foster, the science is exposed. We see the layers of an ancient animal peeled back and brought to life by brightly glowing minerals that have replaced bone or tissue, depending on the environment where animate met inanimate. As I admired these photographs from our gallery's walls, I was reminded of the thrill I had as a young boy when I held an inexpensive magnifying glass up to a common piece of granite and saw nature sparkling back at me. Nature still sparkles in my mind.

References

1. *Time Exposed*, art exhibition at the American Center for Physics, College Park, MD 20740, April 18-October 10, 2014, with photographer Norman Barker, painter Christine Gray, and sculptor Esther Ruiz
2. N.J. Barker, C.A. Iacobuzio-Donahue, *The Hidden Beauty in Biomedical Imaging*, *Journal of Visual Communication in Medicine* **38**, **3**–4, 220-227 (2015)

43

Science, Art, and Theater

In August of 1609, a 45-year-old Italian astronomer, physicist, and mathematician, Galileo Galilei, shocked a group of Venetian lawmakers by demonstrating the remarkable capabilities of the telescope. By the end of 1610, Galileo had discovered mountains and craters on the face of our moon, several of Jupiter's moons, and an entire universe of stars invisible to the naked eye. In the process, he turned human understanding of the cosmos on its head. Some 350 miles up the road in Prague, a young German astronomer and mathematician by the name of Johannes Kepler was witnessing the culmination of more than 10 years of detailed and rigorous scholarship—the publication of *Astronomia Nova*, his groundbreaking treatise on the orbit of Mars and the theory of a heliocentric solar system. In honor of these revolutionary events and coinciding with their 400th anniversary, the International Astronomical Union and UNESCO declared 2009 the International Year of Astronomy (IYA) [1], setting an inspiring goal: "to help the citizens of the world rediscover their place in the Universe through the day- and night-time sky, and thereby engage a personal sense of wonder and discovery".

© Springer Nature Switzerland AG 2020
H. Frederick. Dylla, *Scientific Journeys*,
https://doi.org/10.1007/978-3-030-55800-0_43

Fig. 43.1 Whirlpool Galaxy courtesy of www.fromearthtotheuniverse.org. *Credit* S. Beckwith for the NASA/ESA Hubble heritage team

With the anniversary of these events in mind, the American Center for Physics (ACP) hosted the art exhibit *Cosmic Curiosities* [2] to celebrate the IYA by noting astronomy's inspiration to the arts. The exhibition's curator Sarah Tanguy noted that the exhibit celebrates the stargazer and the power of the imagination and creativity to translate the faraway into something tangible that all can appreciate. The exhibition presented the work of two artists who are inspired by the wonders on display in the dark night sky. With a 40-year career as a glass artist [3], Josh Simpson creates dazzling, space-inspired glass planets and platters. Gay Glading is equally passionate about the night sky and astronomy, and uses the complex interplay of science, mythology, and history to inspire the paintings in her *Constellations* series.

It was most appropriate to have a professional astronomer add his voice and life-long love of the skies to this exhibition. My friend and colleague, Kevin Marvel, the Executive Director of the American Astronomical Society (AAS), gave the exhibition's audience at the opening a grand tour of the beauty of the cosmos by displaying astronomical photographs captured by the Hubble Telescope [4] as part of the exhibit. The AAS and other physics societies used the IYA as an opportunity to take advantage of the public's fascination with images and stories from astronomy for public education and outreach on the value of science. Marvel demonstrated the "Galileoscope", a low-cost telescope developed by AAS to celebrate the year and since sold worldwide for science education [5]. This modest cost instrument has helped hundreds of thousands of people around the world to experience the wonder of the Universe by allowing them to see lunar craters and mountains, four moons circling Jupiter, the phases of Venus, Saturn's rings, and countless stars that are invisible to the naked eye.

The exhibition's opening reception concluded with a talk by playwright Karen Zacarias about her play, *Legacy of Light*. [6] Zacarias, who wrote *Legacy of Light* while pregnant with her third child, talked about her inspiration for the play, which focuses on eighteenth century French astronomer Emilie du Châtelet and her present-day counterpart, Olivia, enacting their struggle to balance the dreams of scientific immortality with the responsibilities of motherhood.

During my tenure with American Institute of Physics, I had the pleasure of hosting art exhibitions such as *Cosmic Curiosities* twice a year at the American Center of Physics. These exhibitions were a perfect mash-up of science and art—entertaining to view but also useful for informing the public of the value and enjoyment of the intersection of these two human endeavors. The success of these ventures was due to the skill and keen eye of the series' curator, Sarah Tanguy [7], who personally selected the artists who were inspired by science, or the scientist-artists who have mastered both crafts since the series was launched in 1997.

References

1. The International Year of Astronomy, 2009; https://www.astronomy2009.org/general/
2. *Cosmic Curiosities*, art exhibition at the American Center for Physics (ACP), College Park, MD 20740, April 21-October 16, 2009
3. Josh Simpson, glass artist; https://www.megaplanet.com
4. The text and images related to the astronomy photographs displayed at the 2009 ACP Exhibition are adapted from www.fromearthtotheuniverse.org where additional images and information can be found
5. The history and availability of this unique low cost but useful telescope is described at: https://galileoscope.org
6. *Legacy of Light*, a play by Karen Zacarias; https://variety.com/2009/legit/reviews/legacy-of-light-1200474861/
7. Sarah Tanguy is the guest curator at the American Center for Physics biannual art exhibitions, in addition to being curator of Washington DC's "*Art in the Embassies*" programs; http://www.sarahtanguy.com

44

In and Out of the Plane

If you visit Washington, DC, you will find the usual government build-ings expected for a national capital, but it is also a city of superb museums. The Phillips Collection, a small museum, built in the home of its founder, Duncan Phillips, is the U.S.'s first museum dedicated to modern art. It is a gem not to be missed.

As a scientist, I have always been attracted to the cubist style of paintings introduced by Pablo Picasso and Georges Braque in 1909–1910. Cubism's intersecting planes and lines rearrange an ordinary scene, inviting observers to wander and explore, to take a painting apart and put it back together, and to admire the whole from different perspectives. A simple shape is not so simple. An image on flat canvas can fly off into the third dimension.

Picasso immersed himself in cubism for a time and went on to other artistic ventures. Braque, for the most part, made cubism his signature style throughout the rest of his life. He survived two world wars and painted nearly up to his death in 1963. Duncan Phillips was an important fan and patron of Braque, buying his first Braque painting in 1927 and adding nine others to his collection.

During the summer of 2013, the Phillips Collection hosted a special exhibit, "*Georges Braque and Cubist Still Life, 1928–1945*", featuring 44 paintings, all still-life's [1]. But there is nothing still about these paintings. They focus on very few objects—a vase, a mandolin, an occasional flower, etc.—but these are just props for this artist. In Braque's words:

© Springer Nature Switzerland AG 2020
H. Frederick. Dylla, *Scientific Journeys*,
https://doi.org/10.1007/978-3-030-55800-0_44

For me, no object can be tied down to any one sort of reality … Objects do not exist for me except in so far as a rapport exists between them. It is this "in-between" that is the real subject of my pictures.

Fig. 44.1 *Studio with Black Vase*, 1938, George Braque. *Credit* The Kreeger Museum, Washington, DC, copyright by the Artists Rights Society (ARS), New York/ADAGP, Paris

Given my profession as a scientist and proclivity as a tinkerer, I appreciate the fact that Braque felt compelled to make all of his own paints. He also experimented with various additives and finishes to add texture and varying tone to the finished pigments. For example, he added grains of sand or ground glass for reflectivity. He never covered the entire surface of his finished painting with the usual varnish layer used on most oil paintings. Instead, he might add beeswax or varnish to just one pigment to allow that surface to pop out from the canvas with a satin finish. He was painstaking, sometimes taking two years to finish a painting, with meticulous sketching before his brush hit the canvas. Paint-overs and paint removals were also part of his process.

For those interested in the archaeology of Braque's paintings, this Phillips exhibit included a section that shows some of his paintings illuminated with infrared or ultraviolet, or dissected into elemental components with x-ray fluorescence spectra. This is a great example of science working for art.

To add to the enjoyment of the intersection of geometry and art, The Phillips Collection also featured a companion exhibit by Ellsworth Kelly. At age 90, Kelly was awarded the National Medal of Art by President Obama in 2013.

Fig. 44.2 *Yellow Relief over Red*, 2004, by Ellsworth Kelly. *Credit* Photo by Jerry L. Thompson. ©Ellsworth Kelly Foundation, Courtesy Matthew Marks Gallery

Reference

1. *Georges Braque and Cubist Still Life, 1928-1945*, Exhibition at the Phillips Collection, Washington, DC, June 8-September 1, 2013; https://www.phillips collection.org/events/2013-06-08-exhibition-braque

45

Following the White Line

The genre of art I find most appealing is the Modernists who experimented with bold shapes and exaggerated colors on both sides of the Atlantic during the first half of the twentieth century. It is not unusual to find my scientific colleagues sharing my admiration for an artist such as Wassily Kandinsky with his ingenious incorporation of geometry in his abstract paintings [1] and his respectable color theory studies, which he incorporated in his teaching at the Bauhaus School in Germany. I have long admired Arthur Wesley Dow [2], a renowned artist and respected teacher who brought the Japanese concept of "*notan*"—the interplay between light and dark elements in art—to his students at the Pratt Institute as he emphasized the importance of a full range of tonality and careful use of negative spaces in artwork.

Among Dow's many talents was his mastery of the color woodcut print.

Making a drawing by covering a carved wood block with an ink or dye and then impressing the pattern on to a piece of parchment or paper is one of the earliest known methods of producing an artistic image. There are examples of wood block images from the third century BCE. The earliest methods of incorporating images into hand-inked versions of the Bible used wood-block printing—and the method continued well into the Gutenberg era when moveable type was used for lettering and carved blocks were used for imagery and page decoration.

The wood block print came into its glory during the Edo period in Japan where spectacular full color prints were made depicting village life and landscapes of the seventeenth and eighteenth centuries. The best known of Japanese woodcut artists were Hiroshige [3] and Hokusai [4]. The prints were greatly admired in the Western world. Visitors to Monet's garden-side

© Springer Nature Switzerland AG 2020
H. Frederick. Dylla, *Scientific Journeys*,
https://doi.org/10.1007/978-3-030-55800-0_45

house in Giverny will find the walls covered with these prints. The technique, though admired, had a serious drawback for any would-be practitioner who wanted to learn the craft. For a full color print, individual woodblocks had to be carved for each color and the resulting print was made by carefully inking and registering each block individually over a piece of carefully prepared paper to build up the composite image. The technique was used by Dow for his color woodblock prints and many of his contemporaries in the U.S. and Europe experimented with the technique to broaden their skills and as an homage to the Edo period artists.

An important simplification of the technique was invented at America's first art colony, Provincetown, Massachusetts, during the winter of 1915–1916. Founded in 1899 by Charles Hawthorne, the Provincetown Art School took advantage of Provincetown's unique location at the tip of Cape Cod. The town is surrounded by the beauty of dunes and water and bathed with a "northern Light" that artists crave for natural illumination. The school, from its early beginnings, attracted the best of American artists, usually for a summer visit—the rest of the year was spent in their home cities in the U.S. with as many sojourns for studies in Paris as could be arranged. This routine was interrupted with the outbreak of World War 1, and a group of artists found themselves spending a cold winter in Provincetown in 1915. The leader of this winter group was the Swedish artist, B. J. O. Nordfeldt, and included six women artists: Juliette Nichols, Edna Hopkins, Ethel Mars, Maud Squire, Ada Gilmore, and Dolly McMillan. All of these artists had some familiarity with various forms of printmaking, including the Japanese color multi-block technique. Perhaps it was the unusual isolation of a bleak and cold Provincetown winter, but something in the air spawned the invention of a great simplification to color wood block printing.

Although largely a group effort, Ada Gilmore credits Nordfeldt with inventing what is now called the white-line wood block technique. This new technique freed the art form from the use of multiple carved wood blocks to create a final image. On a single block of wood, each distinct shape of the intended scene is outlined with a carved line on the block. When the carving is complete, a suitable of piece paper is pinned to one side of the block. Individual areas are then lightly painted with watercolor paints or inks. Each wetted area is impressed upon the paper by folding it over the block and rubbing the back of the paper. The process is repeated until the full image is transferred. As in traditional watercolor paintings, layers of paint can be added to the image to add texture and a range of tones. The process sounds laborious, but it is a great simplification over the carving and registration of individual blocks for each color used in the final print.

Art historian, David Action noted that "*immediately, wood block printing was revolutionized. Being able to see the complete picture on one piece of wood, like a painting on a canvas gave new possibilities for creative work in that medium*" [5].

The first prints from Nordfeldt's group were exhibited in NYC in May 1916 and again in Provincetown that summer. Others joined the group in 1916, including Oliver Chaffee and his student Blanche Lazzell. The latter became the most famous of the genre that is now known as the "Province town Print". The local popularity of the technique in Provincetown peaked after the end of WW1 with a group gallery and several exhibitions in 1918. Except for Lazzell and one of her students, Ferol Sibley Warthen, there were very few practitioners of the technique or any notoriety of the art form for the next half century. This is partially due to poor recognition of Lazzell's talents, although she had a prodigious output. Lazzell went back to Paris in the early 1920s, experimenting with cubism, and is now credited as one of first American artists to bring the technique back to the Americas. She spent most of the remainder of her life in Provincetown. By the time she died in 1955, she had made over 500 prints from more than 120 woodblocks [6].

Fig. 45.1 *The Seine Boat* (1927), a white-line woodcut print by Blanche Lazzell. *Credit* Smithsonian American Art Museum

A contemporary print maker from Provincetown, Bill Evaul [7], is given much credit for the technique's revival in the late 1970s and early 1980s. Evaul was introduced to Ferol Sibley Warthen, the last of the living Provincetown Printers, in 1979, while he was continuing his study of traditional print making and was just beginning to experiment with the Provincetown Print. Evaul became enamored with the technique and was able to incorporate his own expertise in traditional printing to expand the white line technique to sizes greater than a meter in length. Evaul and Ferol Sibley Warthen's granddaughter, Katheryn Smith [8], have been the vanguard of the third generation of Provincetown Printers who have revived this unique art form. A key exhibition debuted at the Provincetown Art Association and Museum in 1982, which then travelled to Washington's National Gallery of Art in 1983, put the technique back into the artistic lexicon.

I missed both of these exhibitions—largely because I was solely dedicated to physics at the time. When I visited a Boston Museum of Fine Arts exhibition of the Provincetown Printers in 2001, I realized I had to make a trip to Provincetown to see what all the fuss was about. With my first visit to a Provincetown art gallery, I was delighted to find an original Blanche Lazzell print, but greatly disappointed to find that the price way exceeded my budget. By chance I picked up a copy of a local art magazine whose cover story featured Bill Evaul who was making contemporary white line woodcuts. I met Bill at his studio the next day and was pleasantly surprised with two findings—first, I enjoyed his art and could afford to buy some prints, and second, we recognized each other. Fifty years earlier we were both members of the same Boy Scout troop and had risen to the top rank of Eagle Scout. Since then we have renewed our friendship. During annual visits to Provincetown, I take lessons from Bill on the white-line technique and I teach Bill physics. The Provincetown Art Association and Museum paid tribute to Bill's four-decade career of producing and promoting the white-line print art form with an exhibition of his work in 2016 [9].

Fig. 45.2 Bill Evaul in his studio in Provincetown with one of the author's woodcuts. *Credit* Author's photo

I find the technique of making white-line woodcuts to be a perfect mixture of art forms I enjoy: wood carving and watercolor painting. The process can't be rushed so its slow deliberate pace gives the brain plenty of Zen-like time for refreshment from the frantic pace and information overload of twenty-first century life. And the resulting print displays the bright colors that any Modernist would love, with an interplay of textures imparted by both the woodblock grain and the printing paper. I have been making prints and enjoying the art form for nearly 20 years. Bill would call me a 4th generation Provincetown Printer, and to commemorate Bill's expertise as an artist and teacher, I have been teaching a 5th generation of students to follow the white line for the last few years [10]. We all hope this art form has fully settled into the fabric of art and art appreciation for the present and future.

References

1. Wassily Kandinsky, *Point and line to plane*, Solomon R. Guggenheim Foundation, New York, 1947
2. Arthur Wesley Dow, *Composition: understanding lines, notan and color*, 9th edn. Doubleday, Page & Co., New York, 1920

3. Mikhail Uspensky, *Hiroshige: One Hundred Views of Edo*, Barnes and Noble Books, New York, 2005
4. Smith, Henry D., *Hokusai: One Hundred Views of Mt. Fuji*, George Braziller, Inc., Publishers, New York 1988; also see https://www.katsushikahokusai.org
5. David Acton, *The Provincetown Print*, in Ref. 6, Chapter 6, pp. 169–206
6. *Blanche Lazzell, The Life and Work of an American Modernist*, Robert Bridges, Kristina Olson, and Janey Snyder, editors, West Virginia University Press, 2004
7. *William Evaul, The Provincetown Printers: Genesis of a Unique Color-Woodcut Process*, *Print Review* **18**, Pratt Graphics Center, New York, 1984
8. Kathryn Smith, *The Provincetown Print,* Provincetown Art Association and Museum, The Permanent Collection, Provincetown, MA, 1999
9. *William Evaul and the White-Line Print Tradition*, 2016, The Provincetown Art Association and Museum Exhibit Catalog for July 1 - August 21, 2016 exhibition of William Evaul's artistic work; also see Bill Evaul's website: www.evaul.com
10. The author's website for white-line art and instruction: see www.freddylla.com

Acknowledgements

A reader of a few essays in this book will observe that I had the good fortune in my career as a physicist to interact and learn from many—my family, teachers and mentors, friends and colleagues, and even from brief encounters with remarkable people who made singular and long-lasting impressions. In attempting to list just some of the people who deserve acknowledgement, I will surely miss deserving individuals. For that faux pas, I ask their forgiveness and thank them for crossing paths with my scientific journey.

I thank my parents, Hank and Frances for putting me on the planet and for indulging my childhood by giving me free range with scientific experiments in electronics, rocketry, chemistry, biology, mineralogy, and astronomy. These early experiments convinced me to become a physicist despite putting certain sections of the household at risk of exposure to high voltage, noxious chemicals, and some low-level radiation. Two high school science teachers, John D. Jones in biology and Fred Schwartz in chemistry, taught me their love of their science and encouraged my science adventures both inside and outside of the classroom. From dedicated essays in this book, you can see how much I gained by spending eight years at MIT with Professors John G. King and Rainer Weiss—from their teaching, coaching, creativity, and incessant scientific curiosity. I was fortunate to be learn from other remarkable members of the physics faculty during my MIT student days: Tony French, Victor Weisskopf, Jerry Friedman, Frances Low, Gene Stanley, Uno Ingard, Phillip Morrison—to acknowledge only a few. During my undergraduate years, the playwright A. R. (Pete) Gurney taught literature courses at MIT. I gained an appreciation of great literature and the ability to write clear and informative sentences from invaluable time spent in Pete's courses.

© Springer Nature Switzerland AG 2020
H. Frederick. Dylla, *Scientific Journeys*,
https://doi.org/10.1007/978-3-030-55800-0

During my years at Princeton University's Plasma Physics Laboratory I was exposed to a cadre of bright and energetic physicists and engineers dedicated to the difficult problem of taming high temperature plasmas for fusion energy. I learned from my exposure to Lab directors Melvin Gottlieb, Harold Furth, and Rush Holt that big science requires scientists to speak plainly and credibly to those who control the flow of public funds for supporting research—the politicians and government administrators. I learned from the project managers for the three big Princeton fusion devices, Wolfgang Stodiek, Dale Meade, and Don Grove, that the scientific method is essential for the constant stream of decisions needed to manage the competing demands of staff, budget, and schedule constraints of a big science project. During my first year at Princeton, I met Fritz Wagner, a visiting postdoctoral fellow from the Institute of Plasma Physics in Garching, Germany. We became lifelong colleagues sharing a love for science, art, and a good sense of humor. Fritz's work on renewable energy is one essay in this volume.

As my mentor at Jefferson Lab, Hermann Grunder, the director who got the lab built, showed me by demonstration how to balance big science project priorities and the value of brief, frequent, and informative communications to all who can make or break a project. I ran two large Jefferson Lab projects: the team that built the superconducting accelerator hardware and a second team that built a unique and powerful laser research facility (both described in this book). All of my colleagues on both teams were essential for both projects' success.

I moved to my last career stop in 2007, becoming the CEO of the American Institute of Physics (AIP). I interacted with hundreds of remarkable AIP colleagues, colleagues in the scholarly publishing community, and members of the U.S. Congress, their staff on the various science committees, and the administrators of the science funding agencies of the U.S. government. It is from my AIP experience that most this volume's essays were originally penned. Special acknowledgement goes to a group who provided essential mentoring during this diverse experience at AIP: Charles Duke (Xerox Research), Ben Snavely (Kodak Research), Lou Lanzerotti (Bell Labs), Robert Doering (Texas Instruments), Robert Campbell (Wiley), John Vaughn (Association of American Universities), Madeleine Jacobs (American Chemical Society), Jim O'Donnell (Arizona State University), Ann Okerson (Yale University), and Scott Plutchak (University of Alabama).

During my entire scientific career, there have been several colleagues with significant impact on my scientific development: Mark Cardillo (Bell Labs), Dennis Manos (College of William and Mary), Joe Greene (University of Illinois), Rox Anderson (Harvard Medical School), Garry McCracken (Culham

Lab, UK), and Paul Redhead (National Research Council, Canada). Their impact affected my scientific worldview.

Despite telling you that many in my professional life have helped me with writing, I have to admit that my writing skills require constant improvement. This volume was immeasurably improved by the writing and editing skills of my wife Linda Dylla (science writer and former head of Jefferson Lab Public Affairs), Liz Caron at AIP, Steve Corneliussen (from Jefferson Lab and AIP), and my editors at Springer, Angela Lahee and Stephen Lyle.

For background research on several topics in this volume, I thank Richard Jones, Jason Bardi, and John Haynes at AIP, and the directors and staff of AIP's Center for the History of Physics, Spencer Weart and Greg Good and their helpful staff at the associated Niels Bohr Library and Archives, particularly the head librarians, Joe Anderson and Melanie Mueller.

Finally, I thank my two daughters, Kim and Sarah. Both grew up to be fine artists and writers, have taught me to enjoy life and see the invaluable connections between art and science, and patiently taught me to improve my nascent skills as a woodcut artist. As I describe in the last essay in this volume, my friend and decorated woodcut artist, Bill Evaul, continues to teach me the joy of the woodcut while I teach him the joy of physics.

Epilogue

As I was completing this book in early 2020, the COVID-19 coronavirus pandemic was enveloping most of the populated areas on Earth. Not since the Spanish Flu pandemic a century earlier has the human population seen such a medical scourge. Only this century, the global connectivity of human travelers and commerce enabled by modern air and sea travel has guaranteed wide dispersal of the virus worldwide in a matter of weeks since the appearance of the outbreak in China. Given this virus's unpleasant characteristics of high transmissivity and often asymptomatic appearance, it spread with deadly consequences, especially in areas of dense population where health systems and general living conditions were poor. Given the lack of a vaccine for this new virus, the only defense was population isolation to slow the spread of the virus. This required most of the world's commerce and travel to grind to a halt, adding to the economic woes, especially for those dependent on daily commerce for food and sustenance.

With the Spanish flu 1918–1919 pandemic, forced isolation of the infected was also the primary means of quelling the infection rates. The medical interventions for the infected at the time–periodic bloodletting and shots of whiskey—were certainly not helpful. Twenty-first century medicine will be essential for quelling the COVID-19 virus. Within a month of its appearance in China, the virus's genome was decoded, allowing it to be tracked as it moved through the world's population, monitored for genetic variations as the virus evolved, and most importantly providing a basis for developing an effective vaccine. Soon after the documented appearance of

© Springer Nature Switzerland AG 2020
H. Frederick. Dylla, *Scientific Journeys*,
https://doi.org/10.1007/978-3-030-55800-0

the virus, virtually all biomedical research establishments in the developed world were studying its make-up, potential therapeutic treatments for the infected, and ultimately effective vaccine candidates. Within two months of the virus's first detection, more than 3000 additional genetic analyses had been performed, a variety of development paths for vaccines described, potential means of accelerating vaccine trials analyzed, and more than 50,000 journal articles associated with COVID-19 were published and made freely accessible to the general public.

This scientific path will have to be followed to quash COVID-19 as a human pathogen, as illustrated by modern medicine's success with prior problematic viral infections like polio in the 1950s and HIV-AIDS in the 1980s. For the world to recover from this pandemic, it may require up to two years or more of scientific research and proper testing of vaccine candidates. Of equal importance to the availability of a potential medical cure is how the world uses the new science and adapts to the economic disruptions caused by the pandemic. The sociology and politics of adapting the world's living conditions and commerce will be as important as the cure in the post-pandemic world.

Dealing with this pandemic and all of its as yet unknown effects on humanity will provide important lessons for life in the twenty-first century. This pandemic is highlighting the inequities in the world's distribution of resources to maintain life and recover from disasters. Will curing the world from the COVID-19 pandemic give us the will to begin alleviating these inequalities? Can the same global cooperation that has led to near instantaneous communications to most corners of the world and widespread manufacture and distribution of products and services be used to more effectively solve global problems such as diseases with known cures, environmental clean-up, and mitigation of climate change? The answer should be yes. This latter question is another manifestation of the two cultures pointed out by C. P. Snow a half century ago and highlighted by Rush Holt in the foreword to this book. Science needs the humanities and vice-versa to solve our most important problems—especially for a pandemic that respects no national borders nor station in life.

There will be important lessons for the world's leaders to learn from the pandemic experience and aftermath. I am optimistic about the world's recovery because science will provide a cure. And most have realized that a full recovery requires worldwide collaboration and cooperation on a scale not seen since the recovery from World War II. May we have another period of post crisis optimism and widespread goodwill. I will do my part as a scientist and a citizen of the world to help move us forward as we address this crisis.

Printed in the United States
By Bookmasters